W9-CSE-930

Oxford Physics Series

Oxford Physics Series

---

1. F. N. H. ROBINSON: *Electromagnetism*
2. G. LANCASTER: *D.c. and a.c. circuits*
3. D. J. E. INGRAM: *Radiation and quantum physics*
4. J. A. R. GRIFFITH: *Interactions of particles*
5. B. R. JENNINGS and V. J. MORRIS: *Atoms in contact*
6. G. K. T. CONN: *Atoms and their structure*
7. R. L. F. BOYD: *Space physics; the study of plasmas in space*
8. J. L. MARTIN: *Basic quantum mechanics*
9. H. M. ROSENBERG: *The solid state*
10. J. G. TAYLOR: *Special relativity*
11. M. PRUTTON: *Surface physics*

M. PRUTTON
Department of Physics, York University

Surface physics

Clarendon Press · Oxford · 1975

Oxford University Press, Ely House, London W.1
GLASGOW NEW YORK TORONTO MELBOURNE WELLINGTON
CAPE TOWN IBADAN NAIROBI DAR ES SALAAM LUSAKA ADDIS ABABA
DELHI BOMBAY CALCUTTA MADRAS KARACHI LAHORE DACCA
KUALA LUMPUR SINGAPORE HONG KONG TOKYO

ISBN 0 19 851819 6

Text set in 9/11 pt. IBM Press Roman, printed by photolithography, and bound in Great Britain at The Pitman Press, Bath

Editor's foreword

Solid-state physics is now a well-established ingredient of the majority of undergraduate physics courses, and it is now possible to give a coherent picture of the chemical composition and structure and of the electrical and thermal properties of condensed matter at a microscopic level. The study of the surfaces of solids often receives little attention in spite of the great importance of the subject for physicists, metallurgists, materials scientists, and electronic engineers. The neglect is to some extent understandable; only in the last decade or so has it been possible to make properly reproducible physical measurements on clean surfaces, and our understanding of the 'surface state' is still much more tenuous than our knowledge of the solid state. Yet it is partly for that reason that surface science is worthy of the attention of science and technology undergraduates; the field is in a state of rapid development, rival theories for various phenomena still abound, and there is the right blend of controversy and consensus to give the student the feeling for what scientific research activity is really like.

Dr. Prutton's book emphasizes the physical aspects of the subject – it reviews the physical techniques that are now available to give reliable diagnostics of crystal surfaces – yet it should also have wide interdisciplinary appeal. Case studies of the application of these techniques to important technological surfaces convey clearly the present state of the art. A general background of solid-state physics or chemistry is a necessary prerequisite, but no sophisticated mathematics is needed. The book is suitable as a text for a final-year undergraduate course in physics, chemistry, materials science, or electrical engineering and is ideal initial reading for anyone setting out to do research in this area or wishing to have a general view of the field.

The solid state by H. M. Rosenberg (OPS 11) provides an ideal base for Dr. Prutton's book, though the more qualitative background given in *Atoms in contact* by B. R. Jennings and V. J. Morris (OPS 5) is sufficient for much of the material.

J.A.D.M.

Preface

The last ten years have seen an enormous growth in many aspects of surface science. Physicists, chemists, and metallurgists have directed their attention to measuring and understanding phenomena at surfaces, all with a view to being able to describe processes which are scientifically interesting or technologically significant. The technological drive required to understand the processes of catalysis and oxidation and to help manufacture semiconductor devices with higher yields has boosted activity in the field enormously. As this book is written this pressure has resulted in the application of a wide variety of new techniques to surface problems but, as yet, it is too early to see whether real progress will be made in understanding such complex processes as catalysis and corrosion. Nevertheless, large strides have been made both in the development of sophisticated techniques with appropriate sensitivities for surface studies and in the initial stages of the understanding the application of these techniques to the measurement of the properties of the surface of simple solids.

This book is based upon a lecture course for final-year Physics undergraduates and new postgraduates. It is intended not to be so comprehensive that the student could read it and go away and start research, but rather to act as a broad introduction to a large subject. Thus, an attempt has been made to describe why surface studies are important scientifically and technologically, what techniques are available and how they compare with each other and with bulk methods, and what types of problems have been and are being tackled. Chapters 2 and 3 are concerned mainly with techniques for the determination of what kinds of atoms are present on a surface and how they are arranged in space in relation to each other. Chapter 4, 5, and 6 then deal largely with selected case studies of particular systems chosen so as to illustrate the application of the techniques to problems concerning respectively the electronic, vibrational, and adsorptive properties of surfaces. The references given to further reading are mainly to other books or to review papers, but it is hoped that this will be sufficient to provide a useful entry to the large literature of surface work.

In order to fit the material into the length available, very many uncomfortable choices have been made. Different kinds of emphasis can be found in G. A. Somorjai's book (*Principles of surface chemistry* (1972), Prentice-Hall, New Jersey, where more weight is given to interactions of gases with surfaces and in J. M. Blakely's book (*Introduction to the properties of crystal surfaces* (1973), Pergamon Press, Oxford), where more attention is devoted to the thermodynamics of surfaces.

I am very pleased to be able to acknowledge the help given me by many colleagues. Drs J. A. D. Matthew, A. Chambers, and T. E. Gallon have read the whole manuscript painstakingly and have contributed many valuable points as well as countering my pedantry. Christine Upton typed the manuscript flawlessly in spite of much other work and constant interruptions. Alan Gebbie drew most of the figures for me.

Colleagues in other laboratories supplied original drawings for some of the figures; I am grateful to Professor W. Roberts, University of Bradford, for Fig. 2.7; Dr. F. Grønlund, University of Copenhagen, for Fig. 3.12; Dr. R. Reid, New University of Ulster, for Fig. 3.14(a); Dr. J. May, Eastman–Kodak Inc., New York, for Fig. 3.14(c); Professor E. W. Müller, Pennsylvania State University, for Fig. 3.19; Dr. D. W. Bassett, Imperial College, for Fig. 6.14; and Dr. H. Montagu-Pollock for Fig. 4.5.

M. PRUTTON

Contents

1. Introduction

All the properties of a piece of bulk material are determined by the number and types of atoms it contains and by their arrangement in space with respect to each other. Some properties can be related in a straightforward manner, both theoretically and experimentally, to the chemical composition and to the crystal structure by using the large body of understanding provided by the band theory of solids (e.g. Rosenberg 1975). Thus, for instance, the division of crystalline solids into insulators, semiconductors, and conductors, the explanation of the relationship between thermal and electrical properties, and the occurrence of both Hall and magneto-resistance effects can all be satisfactorily explained within the framework of the band theory of solids.

It may be less straightforward to relate other properties to a theoretical model of a solid, and a more empirical approach may have to be adopted. One example of such a property is the ferromagnetism of some metals. This depends upon small differences between large interactions in the solid, and demands a difficult and sophisticated theory for its explanation. Mechanical creep and fatigue failure are examples of phenomena requiring an understanding of faults which occur in crystalline solids and the way in which they move in response to applied forces. Again, the theoretical description of these processes is difficult. Nevertheless, these properties are still determined in principle provided that the composition and structure of the material are sufficiently well defined.

The subject of surface physics is the study of the chemical compositions and atomic arrangements at the surfaces of solids and the theory and observation of their mechanical, electronic, and chemical properties. As in the study of bulk solids, the ultimate objective is the establishment of understanding of the relationships between the properties, the composition, and the structure. There are many reasons for expecting that a solid surface will have different properties from the bulk material and these provide an incentive for the physicist to enquire and try to understand. Equally important, there are many processes of technological significance which depend upon the use of solid surfaces and which may be improved in some way if the role of the surface could be fully understood.

In this book the surface is thought of as the top few atomic layers of a solid. In many older books on surface chemistry or metallurgy the surface is regarded as the top 100 nm or so of the solid. The larger distance was determined more by the techniques that were available at that time than by any more basic physical consideration.

The reasons for the expectation that a surface will have different properties from the bulk of the solid may be understood by considering a surface formed by cutting through the solid parallel to a chosen plane of atoms. If the atoms are not disturbed from their bulk equilibrium positions by this operation then the surface can be said to be a *bulk exposed plane* (Fig. 1.1(a)). Such a plane shows the minimum disturbance

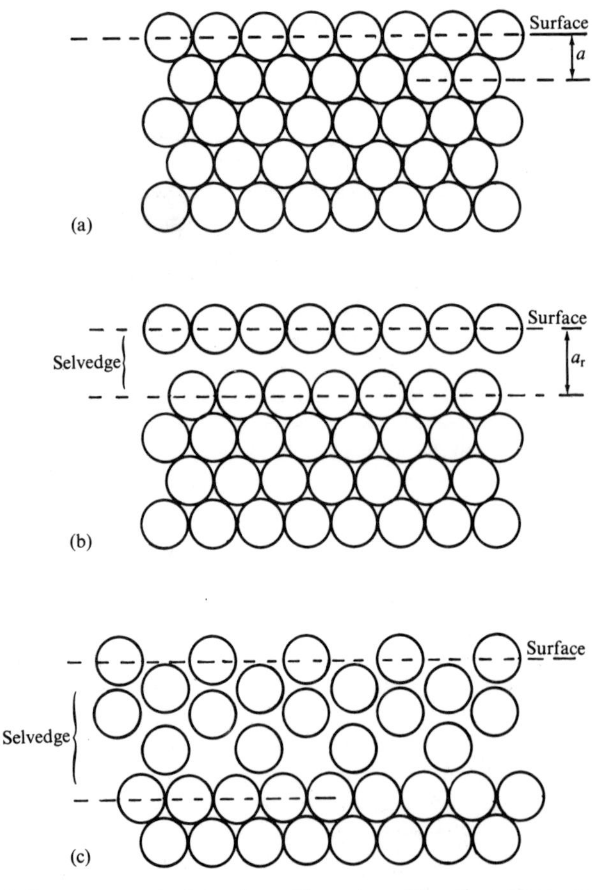

FIG. 1.1. Rearrangement of atomic positions at a solid surface. Hexagonal close-packed atoms. (a) The bulk exposed plane; (b) relaxation of the surface plane outwards; (c) reconstruction (hypothetical) of the outer four atomic planes.

of the solid arising from the formation of the surface. Even so, because many electronic properties of the bulk depend upon the three-dimensional periodicity of the potential inside the solid, the loss of periodicity in one dimension due to the existence of the surface will result in a change in the electronic states near and at the surface and so surface electronic properties different from bulk (Chapter 4). Also, the lack of nearest neighbours on one side of the surface atoms may make available chemical bonds which 'dangle' into the space outside the solid and which will be available for chemical reactions (Chapter 6).

It is more likely that the disturbance caused by terminating the solid in a surface, particularly the disturbance due to the absence of the bonding force of nearest neighbours on one side of the surface atoms will result in new equilibrium positions for the atoms in and near the surface. The simplest change of this kind is the *relaxation* illustrated in Fig. 1.1(b). Here, the separation of the surface plane from the next plane of atoms happens to be drawn so as to be greater than the corresponding separation in the bulk solid. This deviation from the bulk spacing may continue, in decreasing magnitude, as one probes deeper into the solid. The surface region over which there is a deviation from the bulk lattice spacing is referred to as the *selvedge.* Relaxation retains the symmetry of the atomic arrangement parallel to the surface but changes the spacings normal to the surface. It may result in changed properties for the surface because, for instance, it could create an electric dipole moment in the selvedge. A more extreme disturbance occurs when the surface atoms rearrange themselves into a structure with different symmetry altogether different from the bulk solid. This phenomenon is called *reconstruction,* an example of which is shown in Fig. 1.1(c). This reconstruction modifies the symmetry near the surface and will affect all the structure sensitive properties of the surface–the atomic vibrations, and chemical, optical and electronic behaviour.

Many different kinds of processes which are of great scientific and technological interest occur at surfaces. The variety amongst these processes is very large, a fact which accounts for the wide spread of disciplines involved in surface physics. A few of them are listed below in order to give some view of the incentives for surface investigations.

1. *Thermionic emission.* By raising the temperature, sufficient kinetic energy can be imparted to electrons at the top of the conduction band in a metal for them to be ejected from the surface into the vacuum. This process is known as thermionic emission and is important in many electronic devices and most particularly for the source of electrons in oscilloscope tubes and electron microscopes. The number of electrons which can be obtained by thermionic emission is a function not only of the material but also of the presence of chemical contaminants (the cleanli-

ness) of the emitting face and of its crystallographic orientation. The attempt to understand the factors controlling electron emission is an important part of surface physics which is described at greater length in Chapter 4.

2. *Crystal growth.* The development of methods for growing large single crystals of a wide variety of solids has been crucial to the extension of solid-state physics from the simplest band models and to the ability of electronic engineers to design semiconducting devices. The processes of crystal growth generally involves the deposition of atoms upon single-crystal surfaces under such conditions that the arriving atoms can diffuse about and build up the three-dimensional periodic array. Thus, the physics of the energetics and kinetics of adatoms upon single-crystal surfaces is fundamental to an understanding of crystal growth. Some of this type of surface physics is described in Chapter 6.

3. *Chemical reactions.* Many chemical reactions involve interactions between different kinds of atoms across a surface or interface. Even the simplest processes, when viewed at an atomic level, are not fully understood in these terms. One particularly important example is the corrosion of metals or a simple extreme case of corrosion – oxidation. The way in which a clean metal surface is converted to a bulk oxide when exposed to an atmosphere of oxygen must be understood before interpretation of corrosion in more complicated atmospheres is unambiguous. Some simple examples of oxidation on low-index metal faces are described in Chapter 6.

4. *Catalysis.* The presence of surfaces of a particular metal during a chemical reaction can sometimes cause marked increases in the speed of the reaction (Bond 1974). This catalytic action is technologically important but is the subject of a largely empirical literature. It is one of the longer-term aims of surface studies to throw some light upon the way in which complex practical catalytic systems operate, particularly with a view to finding more economic catalysts than metals like platinum.

5. *Colloids.* Micrometre-sized particles of a solid suspended in a liquid – a colloidal suspension – form an interesting and useful chemical system. Many of its special properties arise from the large surface area of the particles and an understanding of its behaviour must rest upon a knowledge of the role of this surface.

6. *Semiconductor interfaces.* Many semiconducting devices depend crucially upon phenomena that occur at a surface or interface. A junction between p-type and n-type material; a junction between a metal oxide and a semiconductor (MOST devices); the junction between a metal contact and a semiconductor – all three involve the formation of a surface and the preparation upon it of another material. The chemistry

and structure of the surface and the way in which these change as the second material is added will affect the electronic properties across the interface. Some of these matters are described in Chapter 4.

7. *Brittle fracture.* Some metals and alloys have enormous mechanical strengths when under continuous load. However, they can often be broken by the sudden application of a much smaller load; this phenomenon is called brittle fracture and it can be quite an embarrassment! It appears to be due to the migration of impurity atoms to the grain boundaries in a solid which become weak regions under impact. The application of surface techniques to the study of this impurity segregation at grain boundaries may help to provide understanding of the problem and may even lead to the discovery of means for inhibiting the diffusion and so preventing brittle fracture in some materials.

Three different factors have played a part in the rise to the current level of interest in surface physics. In the first place, the theory of both the electronic band structure and the chemical bonding in simple bulk solids has been sufficiently successful that theoreticians and experimentalists have been encouraged to attempt to extend the theory. This extension is being explored in two directions simultaneously – the properties of more complicated ionic and molecular bulk solids and the properties of defects in solids. Amongst the defects the most obvious is the occurrence of a two-dimensional surface bounding a three-dimensional periodic structure. Secondly, the technological pressures mentioned above have become more urgent, and, because techniques became available which could throw light upon the relevant problems, interest has grown in trying to use some of the surface physics as it evolves.

The third is technical rather than historical, but it is nevertheless crucial. It is the development, in association with the interest in space research, of techniques for the achievement of ultra-high vacuum (UHV). This level of vacuum is one in which the rate of impingement upon the surface being studied of molecules from the ambient residual atmosphere in the vacuum chamber is negligible in the time required for the observations. The kinetic theory of gases (e.g. Yarwood (1967)) shows that the rate of arrival of N molecules of molecular weight M at a temperature T K upon a square centimetre of surface from ambient atmosphere at a pressure p Torr is given by

$$N = 2{\cdot}89 \times 10^{22}\, p\, (MT)^{-\frac{1}{2}} \text{ molecules cm}^{-2}\, \text{s}^{-1} \qquad (1.1)$$

The unit of pressure in widespread use is the Torr, which corresponds to 1 mm of mercury. In MKS units (1.1) becomes

$$N = 2{\cdot}24 \times 10^{24}\, p\, (MT)^{-\frac{1}{2}} \text{ molecules m}^{-2}\, \text{s}^{-1} \qquad (1.2)$$

with p in newtons per square metre, and T in degrees K.

In a conventional vacuum system using diffusion pumps and elastomer gaskets the pressure is normally about 10^{-6} Torr and (1.1) shows that this corresponds to approximately $3{\cdot}0 \times 10^{14}$ molecules of nitrogen arriving each second on each square centimetre of a surface at room temperature. Since an atomic monolayer corresponds to about 10^{15} atoms cm^{-2} (interatomic distances being about 0·3 nm) such conditions result in nitrogen arrival rates of a monolayer every 3 s assuming that every molecule sticks to the surface. Since many experiments take longer than a few seconds this represents an unacceptable level of contamination of a surface. Ultra-high vacuum is now generally regarded as the region below 10^{-9} Torr. (1.1) shows that, at room temperature, 10^{-10} Torr corresponds to nitrogen arrival rates of $3{\cdot}8 \times 10^{10}$ molecules s^{-1}, or about 1 monolayer in about 8 hours. At this pressure the mean free path between collisions of molecules of the ambient atmosphere would be about 50 000 km.

The techniques required to achieve ultra-high vacua are reviewed in many textbooks (e.g. Redhead 1968; Yarwood 1967). A diagram of the type of UHV system found in many surface physics laboratories is shown in Fig. 1.2, and Fig. 1.3 is a general view of a multiple-technique system in the author's laboratory. The important features of such systems are as follows.

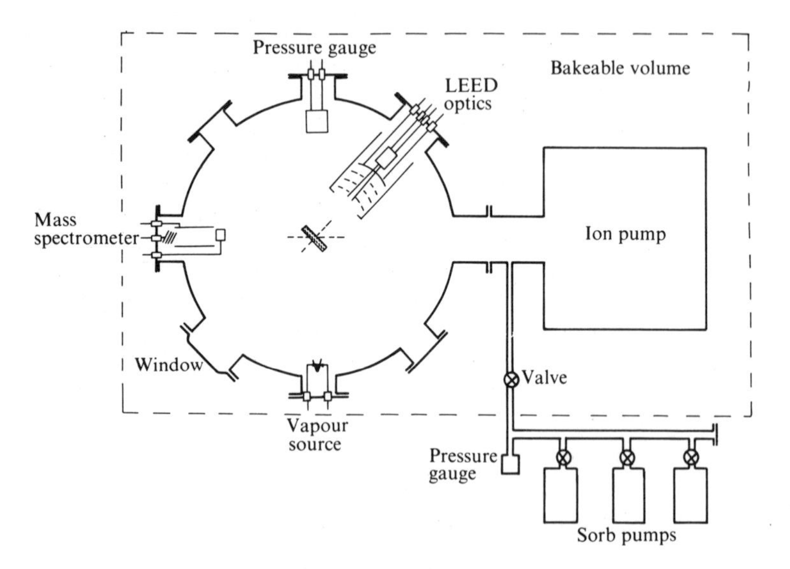

FIG. 1.2. Arrangement of vacuum components in a typical UHV system for surface studies.

FIG. 1.3. A multi-technique UHV system in use for surface studies and equipped for LEED, RHEED, mass spectroscopy, Auger electron spectroscopy, *in situ* optical absorption, and vacuum evaporation.

1. The vacuum chamber and its associated pipework are normally fabricated of argon-arc welded or vacuum-braised stainless steel. This material corrodes very slowly and has low rates of outgassing of absorbed gas.

2. The vacuum joints are made with metal instead of elastomer gaskets. Gold O rings or flat copper rings are normally used here. The use of metal gaskets avoids release of organic contaminants, reduces leakage of water vapour from the atmosphere into the system, and allows the baking described in (3).

3. The whole chamber assembly is designed so that it can be heated to above about 520 K while the vacuum pumps operate. This 'baking' of the system results in accelerated desorption of water vapour (and other gases) from all internal surfaces. When the system is cooled back to room temperature the ultimate pressure attainable is thus substantially reduced.

4. The pumps provided to evacuate the chamber are often ion pumps, titanium sublimation pumps, and, for initial pumping from atmospheric pressure, sorbtion pumps. These three techniques avoid the use of any organic materials. Very well-trapped diffusion pumps filled with special fluids can be used where special problems arise – an important case being to pump away large throughputs of noble gases which are only very slowly removed by ion or titanium sublimation pumps.

5. The choice of materials used inside the vacuum chamber is made carefully to avoid high vapour pressures. Stainless steel, molybdenum, and tantalum are in common use for fabricating parts, oxygen-free high-conductivity copper is often used as a conductor, glass and high-density ceramics such as alumina (Al_2O_3) are used as insulators, and silver as a material to line bearings because it is such a soft metal.

Properly designed UHV systems reach the low 10^{-10} Torr range on a routine basis, and better pressures are possible with great care. Because of the time required to achieve UHV (10 – 24 hours) experiments are often designed to allow several *in situ* operations on the sample surface and several *in situ* types of measurement (Fig. 1.3). The sample itself is normally mounted on some kind of manipulator which may allow various combinations of translation and rotation of the sample in the experimental geometry, electrical isolation, heating, cooling and *in situ* cleavage as a means of preparing a clean single-crystal surface.

Having placed a sample for study in the vacuum system and having achieved a UHV environment for it the next step is often to attempt to obtain an atomically clean surface upon which to conduct experiments. Each crystal face of each material presents its own individual cleaning difficulties, and many man-months of effort can be expended in discovering how to produce a clean surface upon which to start study. This is not the place for a detailed summary of techniques and their applicability but an idea of some of the possible approaches is given in the following list.

1. In some cases the act of achieving UHV environment simultaneously cleans up the surface for study. For example, the cleavage face of mica and the (100) cleavage faces of the alkali halides (NaCl, LiF, NaF, KCl, etc.) can be cleaned in this simple way.

2. As will be seen in Chapter 2, common contaminants on many surfaces are oxygen, carbon, and sulphur which are chemically bound (chemisorbed) to the surface of the material under study. Sometimes these can be removed by heating the sample *in situ*, whereupon the contaminants may desorb into the vacuum as volatile oxides, sulphides, or carbides or may dissolve into the solid leaving a negligible level of contamination behind. Temperatures near to the melting point of the sample material may be required to realize reasonable rates of removal of these contaminants. An example of this method is the removal of oxygen and carbon from the (111) surface of silicon. On heating for 1 or 2 minutes at 1370 K the surface carbon dissolves into the silicon leaving behind sub-monolayer levels of contamination.

3. If the chemisorped contaminants cannot be removed by heating alone they can sometimes be removed by heating in an atmosphere which produces a volatile compound of the contaminant. Thus, surface oxides can sometimes be removed by reduction in a hydrogen atmosphere.

4. More stubborn contaminants can be physically knocked off the surface by bombardment with noble gas ions (e.g. Ar^+, Ne^+) (Redhead 1968). This is usually effective in removing contaminants but is often accompanied by disturbance of the surface atoms of the sample material. For instance, carbon and sulphur contamination can be removed from (100) surfaces of nickel monoxide by bombardment with 200 eV Ar^+ ions, but the nickel monoxide surface is disordered in the bombardment. The surface can be reordered by annealing but this can result in the re-appearance of contaminants by diffusion out of the bulk of the crystal – so-called surface segregation. A combination of bombardment and anneal conditions has to be found empirically which cleans up the surface and allows re-ordering without unacceptable contamination.

5. Some materials cleave naturally on particular crystal faces when struck with a blade in a direction parallel to that face. This property can be exploited to prepare clean surfaces *in situ* and is simple and direct. Examples are (100) alkali halide faces, (111) faces of materials like calcium flouride, and even some materials like silicon and beryllium which will cleave at liquid-nitrogen temperatures. However, this method is limited to a few faces of a small number of materials.

6. Evaporation onto a suitable substrate can be used *in situ* to prepare thin films of polycrystalline or single-crystal type whose clean surfaces can be the subject of subsequent study. The contamination on the substrate can be 'buried' by the deposited film which may then be acceptable for clean surface studies. The growth of oriented single-crystals by this technique is a process known as epitaxy and is described in Chapter 6.

The criteria for deciding if a clean surface had been realized in practice used to be based upon the repeatability of an observation between many samples of the same material prepared in the same way. Although this approach has to be resorted to occasionally it is now more usual to have *in situ* assessment based upon one of the techniques of electron spectroscopy described in Chapter 2.

2. Surface chemical composition

The first questions to be asked about a surface are: what atoms are present and what are their concentrations? If techniques can be found to provide the answers to these questions then the next, more detailed, question is: how are the atoms bound to each other?

The extension of bulk techniques to surface studies

In order to decide whether or not any particular conventional analytical technique will have sufficient sensitivity for useful application to the determination of surface chemical composition it is first necessary to decide what order of magnitude of mass is to be determined. Taking one atomic monolayer of pure aluminium as a sample it is readily calculated that this face-centred cubic (f.c.c.) material must contain about 2×10^{15} atoms cm^{-2}. Since the mass of each atom is about $4{\cdot}5 \times 10^{-23}$ g then a monolayer of aluminium has an areal density of about 10^{-7} g cm^{-2}. A determination of such a monolayer to 1 per cent accuracy therefore requires a technique capable of measuring 10^{-9} g for each square centimetre of monolayer available. Using this figure as a yardstick, a number of well-established techniques are compared in Table 2.1.

The analytical methods of bulk chemical analysis have been highly developed to improve their sensitivities largely because of the demands of thin-film technology and the semiconductor industry (e.g. Maissel and Chang 1970). The microchemical techniques require sufficient sample material to prepare a 1–10 ml solution with approximately 100 μg of sample. Thus, by the yardstick chosen above 10^3 monolayers cm^{-2} are required. This demand arises from the loss of solvent by evaporation during analysis and by loss of solution during transfer from one container to another.

In colorimetry the sample is dissolved off its substrate, and ions in this solution are reacted with specially selected compounds in order to produce a complex with a distinctive absorption spectrum in the visible or near-ultraviolet. Use is then made of Beer's law which states that, for dilute solutions, the optical absorption is proportional to the concentration of the absorbing ion. The method has to be calibrated by using standard solutions and reagents are available to produce suitable complexes with most elements in the periodic table.

TABLE 2.1

Some techniques for chemical analysis in bulk

Technique	Physical basis	Approximate sensitivity (monolayers cm^{-2})	Approximate minimum amount required (g)	Destructive or not	Physical limit to sensitivity or limitations.
Solutions Volumetry	Titration	10	10^{-4}	D	Reagent purity. Adsorption on laboratory ware surfaces.
Spectrophotometry (Colorimetry)	Optical absorption	3	10^{-4}	D	Suitable complexes required. Reagent purity.
Polarography	Current–voltage relations	10	10^{-5}	D	As volumetry.
Flame spectrometry	Emission spectrum	10^{-1}	10^{-4}	D	Optical-detector sensitivity. Noise in fluctuating emission.
Solids Mass Specroscopy	Mass to charge ratio (m/e)	10^{-6}	10^{-12}	D	Ion-detector sensitivity.
X-ray emission	Stimulation of characteristic X-rays	10	10^{-4}	ND	X-ray detector sensitivity. Background of 'white' X-rays.

In flame spectrometry the solution is aspirated into a high-temperature flame and either the emitted radiation is analysed with a spectrometer and the intensity of a selected wavelength is measured or the radiation from a discharge tube containing the element to be determined is absorbed in the flame and the amount of absorption is measured. Again the method is calibrated against standards and sensitivities of sub-monolayer order are possible.

If it is desirable not to dissolve the sample away then techniques using the solid are required. The most common are mass spectroscopy and the X-ray emission spectroscopy, the former being overwhelmingly the most sensitive of established chemical methods. For thin-film and surface applications of mass spectroscopy, atoms are desorbed from a surface thermally by heating the solid near to or slightly above its melting point or by bombarding the surface with electrons, atoms, or ions. The ejected atoms pass into the mass spectrometer, where they are first ionized by electron bombardment. This process can be carried out with very high efficiencies and the ions thus generated passed into a mass filter which, for particular combinations of magnetic and/or electric fields, allows only ions of a particular mass to charge ratio (m/e) to reach a detector. If the detector is an electron multiplier then one ion reaching the detector can result in about 10^6 electrons leaving the detector. It is the combination of high efficiency of ionization, good resolution of m/e, and high gain in the detector that results in the high performance of mass spectrometers and their widespread application. The disadvantage of using mass-spectroscopic techniques are associated with controlling the desorption of atoms from the surface when heating the sample or bombarding it with ions.

Perhaps the most widely used mass spectrometer is the quadrupole type illustrated in Fig. 2.1 (e.g. Redhead 1968). Here, ions generated at position S by electrons from the filament F pass through an aperture A into a region containing four accurately parallel round conducting rods R. The electric field E in the region between the rods has a steady component E_0 and a component $E_1 \cos \omega t$ oscillating at radio frequencies ω. For given geometry and particular values of E_0, E_1, and ω only ions of a particular m/e can oscillate in stable orbits and reach the detector surface D which is the first dynode of an electron multiplier. Electrons generated by the arriving ion are accelerated down the multiplier, generating more electrons as they strike successive dynodes. The signal corresponding to the particular m/e value finally emerges as a pulse of electrons at O for each ion arriving at D. If the rate of arrival of ions at D is sufficiently high the signal at O is a steady current which can be measured with a sensitive amplifier. The mass spectrum can be explored by sweeping through a range of frequencies ω' and the resolving power

$m/\Delta m$, where Δm is the smallest observable mass difference, can be varied by altering the ratio of E_1 to E_0.

In X-ray emission spectroscopy the solid sample is bombarded with electrons or high-energy X-rays so that many atomic levels in the solid are ionized. The atoms return to their ground states by emitting characteristic X-rays. The emitted beams pass into an X-ray spectrometer and the intensity of a selected X-ray line is measured. This is the only non-destructive technique in Table 2.1 but the cross-sections for X-ray emission are rather small and the efficiency of X-ray spectrometers is low so that the sensitivity of the technique is not high enough for analysis of sub-monolayer quantities except under the most favourable circumstances.

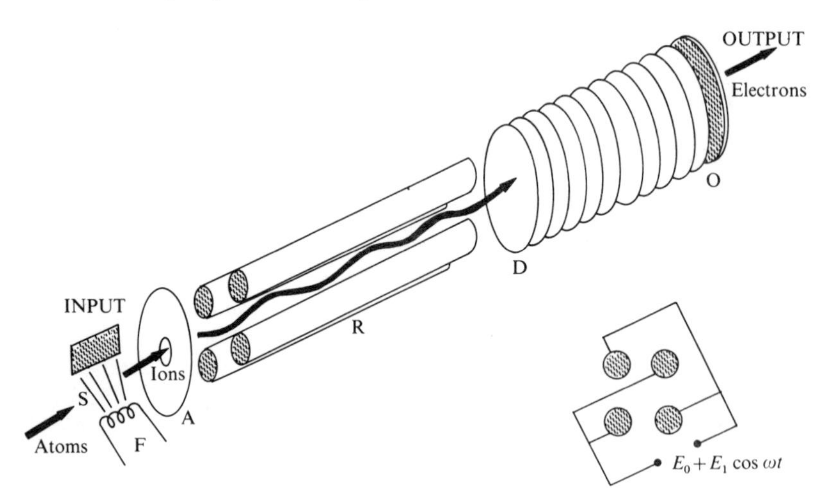

FIG. 2.1. The quadrupole mass spectrometer. For particular values of E_0, E_1, and ω ions of particular m/e pass between the rods on a stable oscillating path. The inset indicates how the rods are electrically connected.

Specifically surface techniques

The destructive character of most microchemical methods and their relative insensitivity in the context of surface studies leads to a search for other methods. The only exception in Table 2.1 is mass spectroscopy which, in spite of its destructive character, is so sensitive that it is of great importance in surface physics, and its application in studies of adsorption and the interaction of atoms or molecules with surfaces is described in Chapter 6. The sensitivity of mass spectrometry is also used in a method called secondary ion mass spectroscopy (or SIMS) which is described on p. 32.

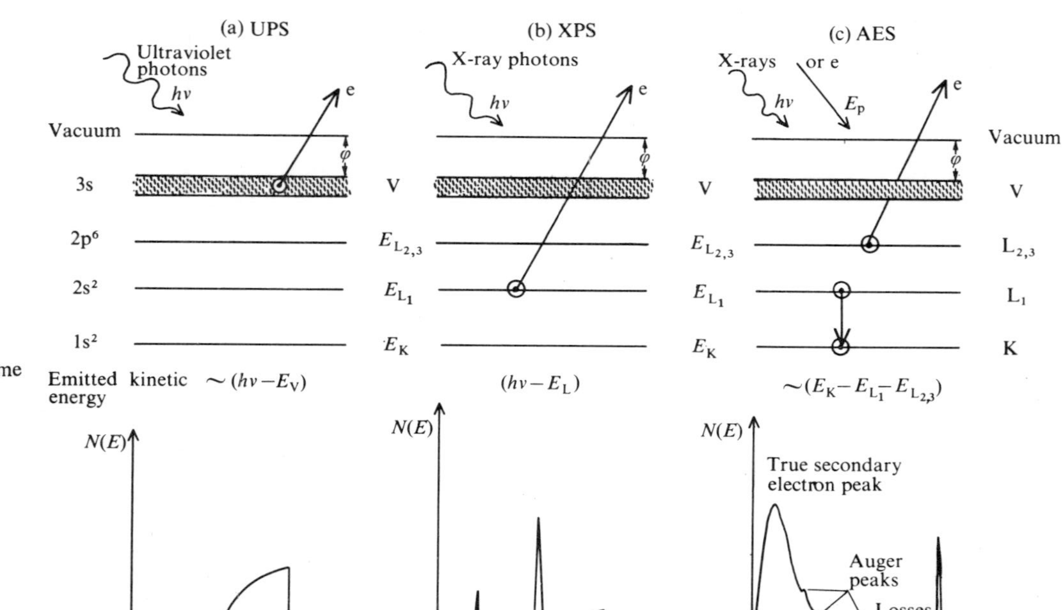

FIG. 2.2. Three kinds of electron spectroscopy. The one-electron energy-level diagrams at the top are for a mythical ideal free-electron metal. They are *not* drawn to scale. At the extreme left are the electronic configurations in which the letters s, p, d, f . . . signify electrons having orbital (angular momentum) quantum numbers 0, 1, 2, 3,. . . ; the numbers to the left of the letter denotes the principal quantum number of one orbit; the superscript to the right of the letter denotes the number of electrons in the orbit. At the extreme right are the labels for the orbitals used in X-ray spectroscopy.

The levels E_K, etc. correspond to the binding energies of the electrons measured from the vacuum level.

The graphs at the bottom indicate the number $N(E)$ of electrons with kinetic energies between E and $E + dE$ which will be measured in the electron energy distribution leaving the solid. The horizontal scales correspond approximately to those that might be found in common practical cases.

The most important methods of surface chemical analysis involve energy analysis of electrons emitted from a surface after it has been bombarded with ultraviolet photons, X-ray photons, or electrons. All these methods of electron spectroscopy use the fact that some of the electrons

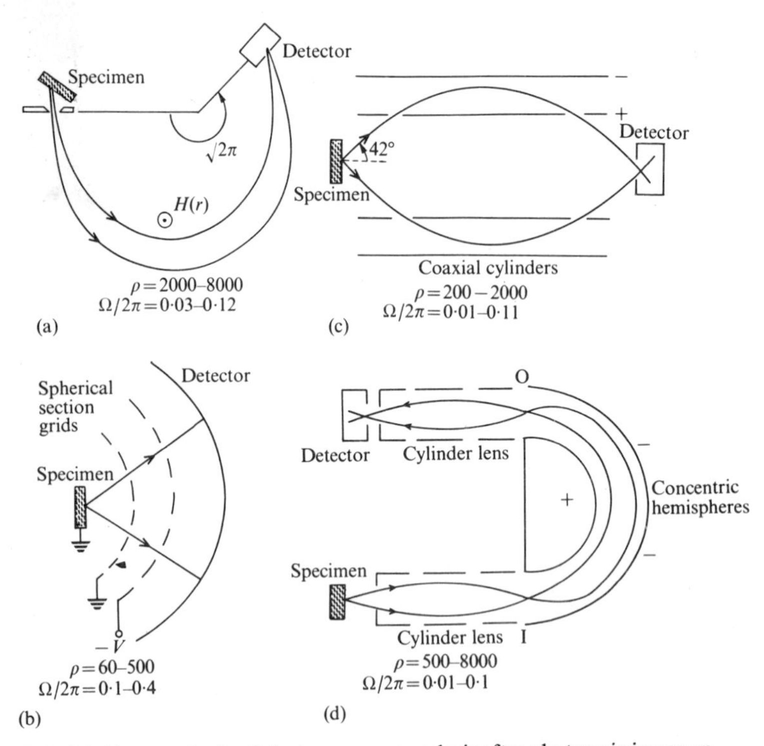

FIG. 2.3. Four methods of electron-energy analysis after photoemission or an Auger process. Typically accessible ranges of ρ and $\Omega/2\pi$ are indicated for each method. (a) The magnetic double-focusing spectrometer. Here the field H is made to be proportional to $r^{-\frac{1}{2}}$ for an electron trajectory of radius r. (b) The electrostatic retarding potential analyser. The potential $-V$ on the second grid results in electrons with kinetic energies greater than eV reaching the detector. The grids and detector are concentric spherical sections with their centre on the sampling surface and at the centre of the region of electron emission. (c) The electrostatic cylindrical mirror analyser (CMA). Appropriate potentials on the outer and inner cylinders result in electrons with a particular kinetic energy being focussed at the detector aperture. (d) The electrostatic concentric hemispherical analyser (CHA). The electron source on the sample surface is focused onto the entrance aperture I of the analyser. With appropriate potentials on the inner and outer hemispheres electrons with a particular kinetic energy are focused on the output aperture O. The three-element output lens then focuses this image onto the (real) aperture of the detector.

emitted have energies characteristic of particular combinations of the energy levels in the solid and so are characteristic of the types of atoms contained in the solid. The processes involved in ultraviolet photoelectron spectroscopy (UPS), X-ray photoelectron spectroscopy (XPS), and Auger electron spectroscopy (AES) are described using one-electron energy level diagrams in Fig. 2.2.

Many techniques are used for the energy analysis of the electrons leaving the specimen. Four common methods are illustrated in Fig. 2.3 and are described in more detail by Sevier (1972). The two most important characteristics of an electron-energy analyser are its energy resolving power ρ and the fractional solid angle $\Omega/2\pi$ of electrons which it will accept for analysis. The resolving power ρ is determined by the slit widths and geometrical factors of the design. It is related to the energy of the electrons E and the spread of energies ΔE_A passed by the analyser through the equation.

$$\rho = \frac{E}{\Delta E_A}. \tag{2.1}$$

Most systems operate at constant resolving power as the analyser is swept through the range of energies under study. The sensitivity of the analyser is determined by the fraction $\Omega/2\pi$, since 2π is total solid angle for backscattering. In most analysers, increased sensitivity can be obtained only at the expense of reduced resolving power.

The highest-precision instruments have been set up and operated by Siegbahn and his co-workers (1967), but these suffer from the disadvantage of being physically large conventional vacuum instruments dedicated to electron spectrometry only. The most commonly used analyser is that of Fig. 2.3(b) because it requires the same electron optics as that used in low-energy electron diffraction LEED (Chapter 3) and is therefore more flexible in its applications. In general, the retarding potential analysers are used for identification of what elements are present upon a surface and simple quantitative studies and the more powerful types of Fig. 2.3(a) and (d) for detailed study of the electron spectrum. The cylindrical mirror analyser Fig. 2.3(b) has good resolving power and high collection of efficiency combined with reasonable physical size and is useful for fast collection of results.

Photoelectron spectroscopy

The photoelectron spectra of very large numbers of atoms and molecules have been studied intensively by Siegbahn and his co-workers (1967), and this work provides an important base upon which to start investigation of solids and solid surfaces. XPS is very often referred to as electron spectroscopy for chemical analysis (ESCA).

In both types of photoelectron spectroscopy, if the incident photon has sufficient energy $h\nu$ it is able to ionize an electronic shell and an electron which was bound to the solid with energy E_B is ejected into the vacuum with kinetic energy E_k. By conservation of energy,

$$E_k = h\nu - E_B, \tag{2.2}$$

neglecting the very small recoil energy of the emitting atom. If the incident radiation is monochromatic and of known energy and if E_k can be measured using an electron energy analyser then the binding energy E_B can be deduced. In UPS the source of radiation is usually a helium discharge lamp which can be made so as to operate at 21·2 eV and at 40·8 eV. In XPS the radiation is usually obtained from X-ray tubes with aluminium or magnesium anodes which give lines at 1487 eV and 1254 eV respectively. The helium radiation has insufficient energy to eject deep core electrons and is used mostly to study the electrons in the valence band of the solid. On the other hand, the X-rays do have sufficient energy for ionization of core levels in many elements and can be used to study electrons in both band and core states.

The width of a feature observed in a photoelectron spectrum will depend upon the intrinsic width of the level from which the photoelectron is ejected, the width of the incident radiation (because $h\nu$ appears in eqn (2.2), and the window ΔE_A of the analyser. In XPS the width of the K_α† radiation incident upon the sample is usually the controlling instrumental factor and this is about 1–1·4 eV unless an X-ray monochromator is incorporated. In UPS the analyser window is normally the controlling parameter as the width of the radiation from a helium II‡ resonance source is close to 1 meV.

Because the number of photons arriving in the incident beam is small in practical cases the intensity of photoelectron lines is low and individual electrons have to be counted as they arrive at the detector.

Part of the XPS spectrum obtained from gold using magnesium K_α incident photons is shown in Fig. 2.4. This Figure shows the counting rates required, on the vertical scale, and demonstrates clearly the narrow-core-like levels N_6 and N_7 which appear as 1·3–1·4 eV wide, and the

†In X-ray notation K_{α_1} is the line in the X-ray emission spectrum due to an electron in the L_3 sub-shell falling into the K-shell hole; K_{α_2} is the line due to an electron in the L_2 sub-shell falling into the K-shell hole. $K_{\alpha_{3,4}}$ arise from doubly ionized K-shell initial states.

‡In this spectroscopic notation He II is the radiation emitted by He^+ ions in the transition from their first excited state to their ground state.

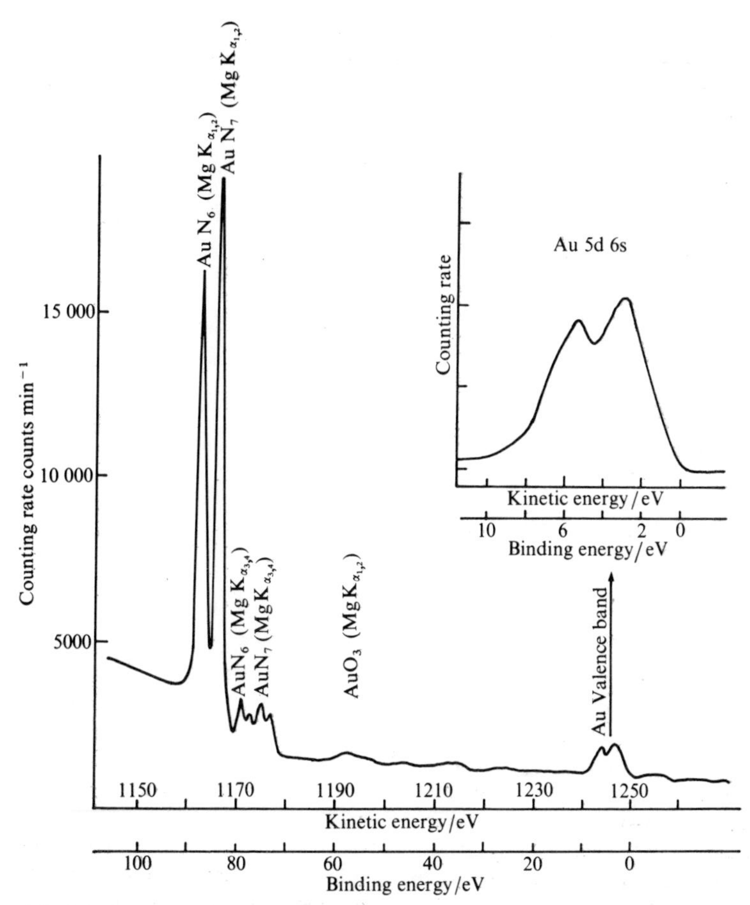

FIG. 2.4. XPS spectrum of the outer levels of solid gold. Both Mg $K_{\alpha_{3,4}}$ and Mg $K_{\alpha_{3,4}}$ X-ray lines are present in the incident beam. The N_6, N_7, and O_3 levels of gold are resolvable as well as the conduction band, which is shown expanded in the inset. The binding energy scale has its zero on the right and is set at the Fermi level. (After Siegbahn *et al.* 1967.)

conduction band which is about 8 eV wide. The peaks are labelled by conventional X-ray notation where the letters K, L, M, N, O . . ., etc. are assigned to orbitals with principal quantum numbers n = 1, 2, 3, 4, 5 . . ., etc. The suffixes give the orbital and total angular momenta quantum numbers l and j of the hole left after photoemission according to the convention in Table 2.2. The X-ray notation is related to the spectroscopic notation in columns 4 and 6 of Table 2.2.

TABLE 2.2 *Atomic and X-ray notations*

Quantum numbers			Atomic notation	X-ray suffix	X-ray notation
n	l	j			
1	0	$\frac{1}{2}$	1s	1	K
2	0	$\frac{1}{2}$	2s	1	L_1
2	1	$\frac{1}{2}$	$2p_{\frac{1}{2}}$	2	$L_{2,3}$
2	1	$\frac{3}{2}$	$2p_{\frac{3}{2}}$	3	
3	0	$\frac{1}{2}$	3s	1	M_1
3	1	$\frac{1}{2}$	$3p_{\frac{1}{2}}$	2	$M_{2,3}$
3	1	$\frac{3}{2}$	$3p_{\frac{3}{2}}$	3	
3	2	$\frac{3}{2}$	$3d_{\frac{3}{3}}$	4	$M_{4,5}$
3	2	$\frac{5}{2}$	$3d_{\frac{5}{2}}$	5	
4	0	$\frac{1}{2}$	4s	1	N_1
4	1	$\frac{1}{2}$	$4p_{\frac{1}{2}}$	2	$N_{2,3}$
4	1	$\frac{3}{2}$	$4p_{\frac{3}{2}}$	3	
4	2	$\frac{3}{2}$	$4d_{\frac{3}{2}}$	4	$N_{4,5}$
4	2	$\frac{5}{2}$	$4f_{\frac{5}{2}}$	5	
4	3	$\frac{5}{2}$	$4f_{\frac{5}{2}}$	6	$N_{6,7}$
4	3	$\frac{7}{2}$	$4f_{\frac{7}{2}}$	7	

(See e.g. Kuhn 1969.)

First-principles theoretical calculations of the energies of photoelectron lines from atoms can be made under two extreme kinds of approximation illustrated by the configurational diagram of Fig. 2.5. In the *'sudden' approximation* the wavefunctions ψ are assumed to be unchanged during the time that the photoelectron is emitted (and the repulsive force on the outer electrons due to the coulomb interaction with the inner electron is removed) and the binding energy of the photoelectron is just the difference E_{BS} in Fig. 2.5. In the *adiabatic approximation* the wavefunctions relax to their ionic form before the photoelectron has left the region of the atom and the calculated binding energy will correspond to E_{BA} in Fig. 2.5. The discrepancy between E_{BA} and E_{BS} for core levels is found to be greater than the difference between the best theoretical value and observation. E_{BA} is the best fit to observation, always falling 1 – 12 eV above the observed binding energy for K-shell calculations on elements with atomic numbers Z between 6 and 23 and for which E_K varies from 288 eV to 5469 eV. This difference may be due to the fact that theory is for an isolated atom of the material and the observation is usually on a solid.

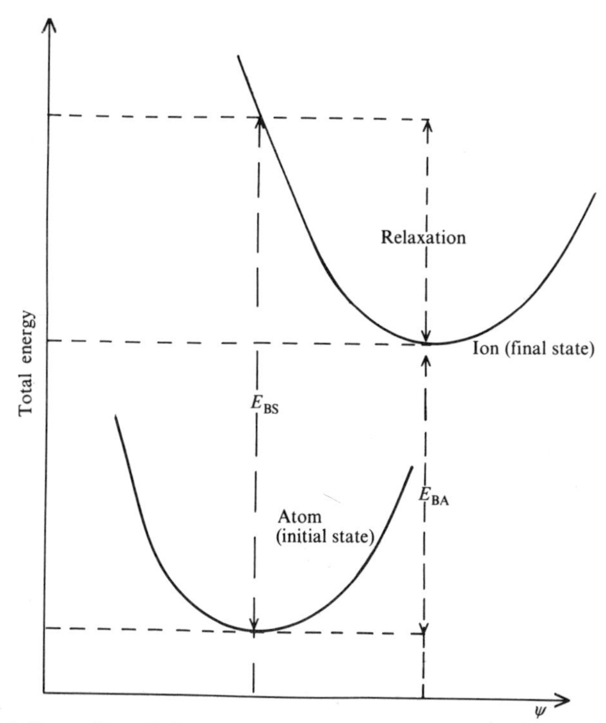

FIG. 2.5. Comparison of the sudden and adiabatic approximations for calculating the binding energy. The total energy of the atom and the ion is shown as a function of the wavefunction ψ.

The valence band spectrum of gold is shown in Fig. 2.6 which was obtained by Spicer (1970) using UPS with a set of helium radiations having $5{\cdot}4$ eV $< h\nu < 21{\cdot}2$ eV. Although the main features of the valence band structure observed by XPS or UPS do agree there is considerably more structure in the data obtained by UPS. This is not only because of the greater energy resolution of the UPS experiment (ΔE in the UPS experiment is about 0·1 eV, in the XPS experiment it is about 1 eV) but also because of differences in the two processes. The interpretation of the UPS spectra is complicated (Spicer 1970; Fadley and Shirley 1970) by the following:

1. The fast variation of the free mean path for electron–electron scattering (Fig. 2.7) at low kinetic energies. Thus, for fixing photon energy, electrons originating from the bottom of the conduction band have a greater mean free path than those originating from near the Fermi level.

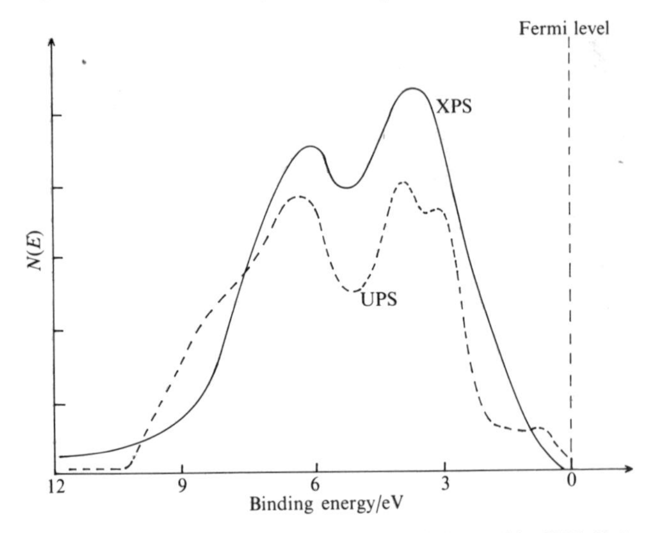

FIG. 2.6. The one-electron density of states for gold obtained by UPS (Spicer 1970) compared with that obtained by Siegbahn (1967).

2. If an electron is ejected with low kinetic energy the probability of photoemission will be affected by the density of states available to the photoelectron. A variation of this empty density of states with energy will affect the shape of the observed *N(E)* curve for the photoelectrons.

The surface sensitivity of the photoelectron methods does not depend upon the penetrating depth of the incident radiation but rather upon the probability that a photoelectron, once generated, will be able to escape to the surface without further energy loss. Electrons can lose energy in a number of ways listed below, each of which is a quantized process.

1. The smallest energy losses, of the order of a few tens of millielectronvolts, are due to the excitation of lattice vibrations or *phonons.* These losses are so small that they are usually detected only in UPS experiments using high-resolution spectrometers. In other cases they are normally part of the low-energy side of a photoelectron peak.

2. Electron–electron interactions can excite collective density fluctuations in the electron gas in a solid. These fluctuations, known as plasmons, are quantized with energies in the range 5–25 eV, the energy depending upon the density of the electron gas and whether or not they are three-dimensional fluctuations – *bulk plasmons* – or two-dimensional fluctuations at the surface – *surface plasmons.* The subject is

reviewed by Kittel (1967). The energy ΔE_B lost to a bulk plasmon is given by

$$\Delta E_B = n\left(\frac{ne^2}{\epsilon_0 m}\right)^{-\frac{1}{2}} \tag{2.3}$$

in a free-electron gas of density n. The energy ΔE_s lost to a surface plasmon is given by:

$$\Delta E_s = \left(\frac{\Delta E_B}{\sqrt{2}}\right). \tag{2.4}$$

3. Various single- and double-particle excitations can occur. Thus an electron may lose energy by raising a second electron from its ground state to an empty state in the solid (e.g. interband transition or a transition from a core level to an impurity state). Alternatively, it may ionize a level with the ejection of another photoelectron and perhaps leading to the generation of another Auger electron (p. 24).

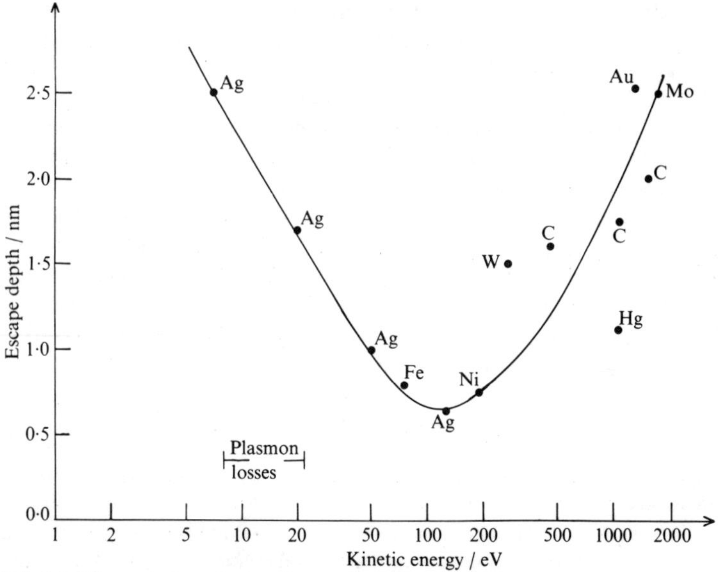

FIG. 2.7. The variation of mean free path prior to an inelastic scattering event of appreciable energy with the kinetic energy of the emerging electron.

Universal curves of this kind should be treated only as a rough guide to the mean free path of the escaping electron. As described in the text, there are several mechanisms responsible for the losses causing this escape depth, and there is no reason to expect that they will add up in such a way that universal curves can be drawn.

(Assembled from data supplied by Professor M. W. Roberts, University of Bradford, and Dr J. C. Tracy, Bell Telephone Laboratory, U.S.A.)

The net effect of these inelastic processes is difficult to calculate, but it is known that in metals and semiconductors the mean free path of an electron prior to inelastic scattering will vary with its kinetic energy. At very low energies the electron will be unable to excite any of above losses and the mean free path will be long. At very high energies, the cross-sections for exciting the losses fall, and again the mean free path will be long. As indicated in Fig. 2.7 the mean free path is believed to pass through a minimum below 1 nm at energy near 100 eV. Even at 1000 eV the mean free path in metals is probably less than 3 nm (about 10 atomic monolayers).

Thus the surface sensitivity of the photoelectron methods will vary with the kinetic energy of the ejected electron but is best in the range 50–200 eV. In UPS using 21 eV radiation the depth sampled will be greater than in using UPS at 41 eV because of the low kinetic energies in the first case. In XPS the surface sensitivity will vary more extremely because of the wide range of kinetic energies possible. Siegbahn (Siegbahn *et al.* 1967) has been able to detect and measure 10^{-8} g of iodine in the $3d_{\frac{5}{2}}$ peak at 620 eV – a figure which corresponds to a sub-monolayer coverage in the terms of Table 2.1 (p. 12).

In summary, the techniques of photoelectron spectroscopy can be used to identify the atoms at surfaces by comparing the lines observed with either calculated core level binding energies or experimentally derived spectra from standards. Quantitative assessments of the amount of each element present are more difficult and discussion of this is deferred until pp. 29–32. Information about bonding of surface species can be obtained by either UPS or XPS by studying the change of shape of valence-band photoelectron features as the surface species is added. Alternatively, the change in energy of a core level (*the chemical shift*) as the environment is changed, can give bonding information.

Auger electron spectroscopy (AES)

As indicated in Fig. 2.2(c), the Auger process is an alternative to X-ray emission and occurs after an atomic level has been ionized by incident photons or electrons. The hole in the inner shell is filled by one electron from a less tightly bound level and a second electron escapes into the vacuum with the remaining kinetic energy. The energy of this Auger electron is very roughly (see later)

$$E \sim E_{\mathrm{K}} - E_{\mathrm{L}_1} - E_{\mathrm{L}_{2,3}} \qquad (2.5)$$

for the transition shown in Fig. 2.2(c), because $E_{\mathrm{K}} - E_{\mathrm{L}_1}$ is the amount of energy released by an electron falling from the L_1 shell to the initial-state hole in the K shell, and the electron escaping uses up $E_{\mathrm{L}_{2,3}}$ of this amount to overcome its own binding energy. Again then, the emitted electrons due to this process will have energies characteristic of the levels of the atoms whence they came and energy analysis will enable identification of the materials present. The Auger electron described by the process of eqn (2.5) would be referred to as a $KL_1 L_{2,3}$ Auger electron. If one or both of the final state holes is in the conduction band of a metallic sample then it is conventional to use the notation V for each of these holes. Thus, an $L_{2,3}$ VV Auger transition would involve an initial state hole in the $L_{2,3}$ shell and two final state holes in the conduction band.

For light elements (atomic number $Z < 20$) Auger emission is more probable than X-ray emission for a K-shell initial-state hole, and for $Z < 15$ it is almost the exclusive process. For higher Z Auger process dominate for initial-state holes in other shells. Thus, if the primary electron beam has an energy below 1000 eV Auger processes will predominate. These high probabilities of Auger emission, coupled with the high flux of incident electrons which is easily achieved in practice, lead to Auger electron spectroscopy (AES) being an extremely sensitive technique for surface chemical analysis.

All the forms of spectrometer shown in Fig. 2.3 are used for AES. To obtain highest sensitivity the incident beam is of electrons arranged to arrive at the surface near grazing incidence by the argument shown under Fig. 2.8. In order to obtain numerical estimates of the current of Auger electrons that will be collected in an experiment Bishop and Rivière (1969) have shown that the equation in the caption to Fig. 2.8 has to be modified as follows.

1. The incident electron current I_0 ionizes atoms on the surface directly but also causes secondary electron emission from underlying atoms. Some of these secondary electrons will also have enough energy to ionize the surface atoms. This effect can be allowed for by incorporating with I_0 a factor r, *the back-scattering factor,* which increases the effective value of I_0.

2. If the cross-section for ionization by an incident electron is Φ then the only means by which the ion can decay to its ground state are by emission of a characteristic X-ray or by emission of an Auger electron. If the proportion of decays attributable to X-ray fluorescence is ω (*the fluorescence yield*) then the Auger cross-section σ_A must be simply $(1 - \omega)\Phi$.

3. In any practical apparatus all the Auger current I_A emitted into 4π st will not be collected but some fraction $\Omega/4\pi$ will arrive at the detector, where Ω is the solid angle accepted by the analyser. This is valid if it is assumed (reasonably) that Auger emission is isotropic.

The nett effect of these modifications is that the total Auger current observed, i_A, will be given by

$$i_A = \Omega I_0 \tau r(1-\omega)\Phi \sec\phi/4\pi. \quad (2.6)$$

Reasonable values for the parameters in eqn (2.6) lead to values of i_A between 10^{-14} and 10^{-12} A, depending upon N, I_0, Ω, and the energy E_p of the incident electrons (which affects Φ). As indicated in Fig. 2.2(c) this current is superimposed upon a background current due to secondary

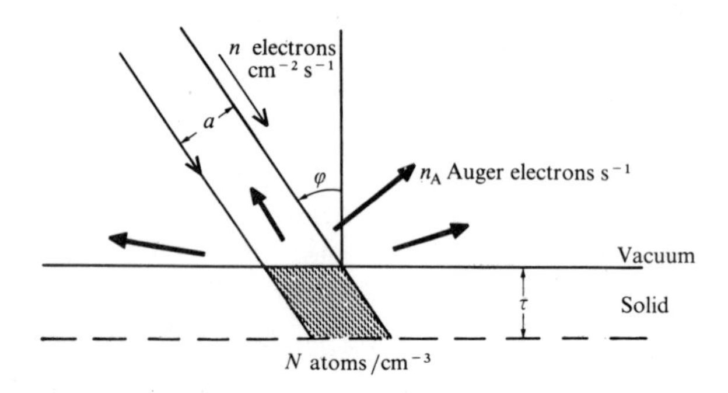

FIG. 2.8. The Auger current generated by a beam current I_0 within an escape depth τ.

Assume: Incident beam has cross-sectional area A. One incident electron causes Y Auger electrons per atom. All Auger electrons generated within τ of surface escape – Auger electrons generated in rest of solid suffer losses in escaping and do not contribute. N atoms cm^{-3} contribute to Auger process in question.

Then: Volume corresponding to hatched area above contains all atoms contributing to Auger signal. Number of contributing atoms $= N\tau A \sec\phi$; nA incident electrons per second give $nA\ YN\tau A \sec\phi$ Auger electrons per second. Auger current $I_A = I_0\ N\tau A \sec\phi$ for incident current I_0. But YA is the cross-section σ_A for Auger emission back into the vacuum.

$$I_A = I_0 N\tau\sigma_A \sec\phi$$

electron emission which carries no information about the type of atoms present and which may be 5 or 10 times larger than i_A. Special electronic techniques can be devised to present measurable Auger signals in the presence of this background but they result in the observation of the differential $d(N(E))/dE$ (or $N'(E)$) of the energy spectrum instead of $N(E)$, which is shown in Fig. 2.2.

An Auger spectrum in the $N'(E)$ form, obtained during a retarding potential analyser of the general type shown in Fig. 2.3(b), is given for two n-type Si (111) surfaces in Fig. 2.9. After careful chemical cleaning in the laboratory the silicon specimen is placed in the spectrometer, and its Auger spectrum is as indicated in Fig. 2.9(a). By using either tables of atomic energy levels (Bearden and Burr 1967) and relation (2.5) or spectra from known materials, the features can be assigned to C, S, O, P, and Si.

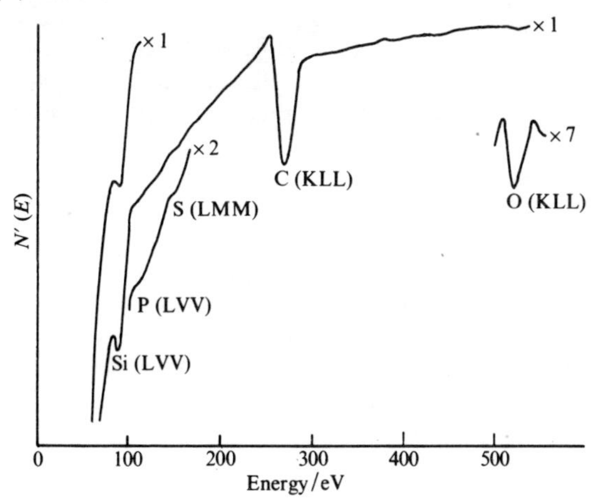

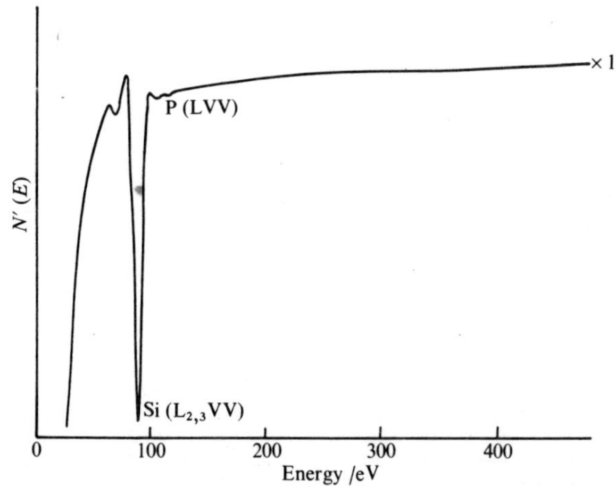

FIG. 2.9. $N'(E)$ Auger spectra from Si(111). (a) The chemically cleaned surface; (b) the 'clean' surface Auger spectrum showing only Si, the dopant P, and sub-monolayer traces of C. (After Gallon, Prutton, and Wray 1971.)

This is a common type of result, the contaminants C, S, Cl, and O being present on many kinds of surfaces chemically cleaned at atmospheric pressure. After cleaning in the UHV chamber by heating near the melting point the contaminants can be removed from the silicon surface and the 'clean' surface spectrum of *n*-type silicon obtained (Fig. 2.9(b)). Careful use of eqn (2.6) indicates that the residual carbon signal in Fig. 2.9(b) corresponds to less than 5 per cent of an atomic monolayer of carbon. This demonstrates the high sensitivity of Auger electron spectroscopy for surface species – a property which arises because of the high ionization efficiency of the incident electrons and the small mean free paths τ for inelastic scattering of the comparatively low-energy Auger electrons.

If a higher resolving power electron-energy analyser is used (e.g. that of Fig. 2.3(d)) then fine structure can be observed in the Auger spectrum. This had been seen many years ago for gaseous samples (Seigbahn 1967) and very similar effects are seen in solids. The $N(E)$ curves for the KLL Auger lines of oxygen and magnesium in the (100) surface of magnesium oxide obtained with an instrument with resolving power of 1200 are shown in Fig. 2.10. The fine structure can be explained in terms of $L-S$

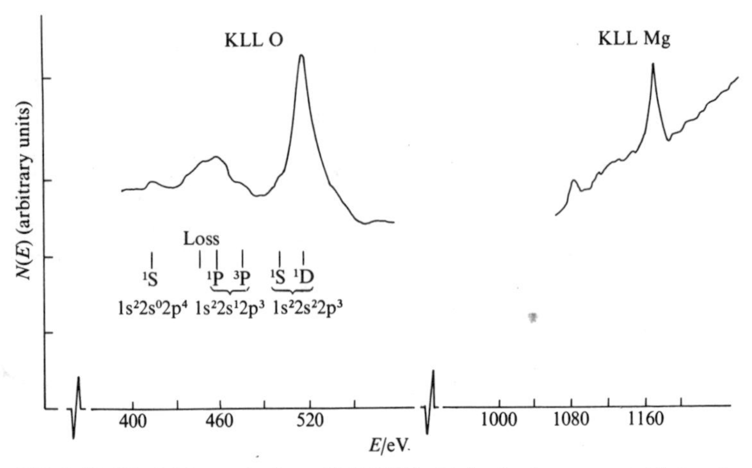

FIG. 2.10. The KLL spectra from MgO (100) obtained using a concentric hemispherical analyser with resolving power 1200. The fine structure is assigned using L–S coupling of the two final-state holes. The final state of the atom (after Auger emission) is indicated using the atomic notation of Table 2.2, the superscripts indicating the number of electrons in each orbital. (After Bassett, Gallon, Matthew, and Prutton 1972).

coupling of two final-state holes left in an atom near the surface after the emission of an Auger electron (for an account of $L-S$ coupling see e.g. Kuhn 1969). Many solids show these quasi-atomic Auger spectra, the features being different from the gas only in that they are all shifted in energy and greater in linewidth.

Careful examination of the energies of observed Auger lines shows that relation (2.5) is not accurately obeyed. The correct energy of an Auger line is not given by differences of the three binding energies as obtained from XPS. This is because of the interactions between the two final-state holes. Referring, as an example, to Fig. 2.2(c), once the L_1 electron has filled the initial-state hole in the K shell then the binding energy of the $L_{2,3}$ electron is increased because the Coulomb repulsion of an L_1 electron has been removed. This hole–hole interaction in the final-state configuration will depend upon whether the holes are both in core shells, one in a core shell and one in a band, or both in a band. Empirical techniques have been devised to allow for this effect, and other complications and these are discussed by Gallon and Matthew (1972). Accuracies of about 5 eV can be obtained by using eqn (2.7) for an Auger transition involving the three levels A, B, and C,

$$E_{\mathrm{ABC}}(Z) \triangleq E_{\mathrm{A}}(Z) - \tfrac{1}{2}\left\{E_{\mathrm{B}}(Z) + E_{\mathrm{B}}(Z+1) + E_{\mathrm{C}}(Z) + E_{\mathrm{C}}(Z+1)\right\}. \tag{2.7}$$

If the two final-state holes are both in the conduction band the electron gas screens the Coulomb interaction between them, and so eqn (2.5) gives a reasonable approximation to energies of AVV Auger features.

In eqn (2.7), E_{A}, E_{B}, and E_{C} are the binding energies of electrons in levels A, B, and C. The use of the binding energies for combinations of the atomic numbers Z and $(Z + 1)$ as indicated makes crude allowance for the hole–hole interaction.

The surface sensitivity of Auger spectroscopy can be measured and eqn (2.6) tested by exploiting the observation that some materials can be grown upon others in such a way that the atoms build up in large monolayer-thick islands layer by layer. That this occurs can be established by some of the techniques described in Chapter 3. An example of such growth is found when silver grows upon the (100) surface of clean nickel, and Fig. 2.11 shows the results of an Auger spectroscopy experiment using this system. As the silver is deposited at a constant rate and the size of the 355 eV silver and 60 eV nickel Auger peaks measured as a function of time. A clear break can be seen when the first monolayer of silver has formed, and it can be seen that 10 per cent of a monolayer of silver could be observed by this method. The experiment can also be used to estimate the mean escape depths of different Auger electrons and these are found to be 0·54 nm for the 355 eV silver electrons and $\sim$ 0·2 nm for the nickel 60 eV electrons.

The quantitative estimation of elements at a surface is difficult with any of the electron spectroscopies in any case, except when the element is uniformly distributed above or if the element is known to be uniformly

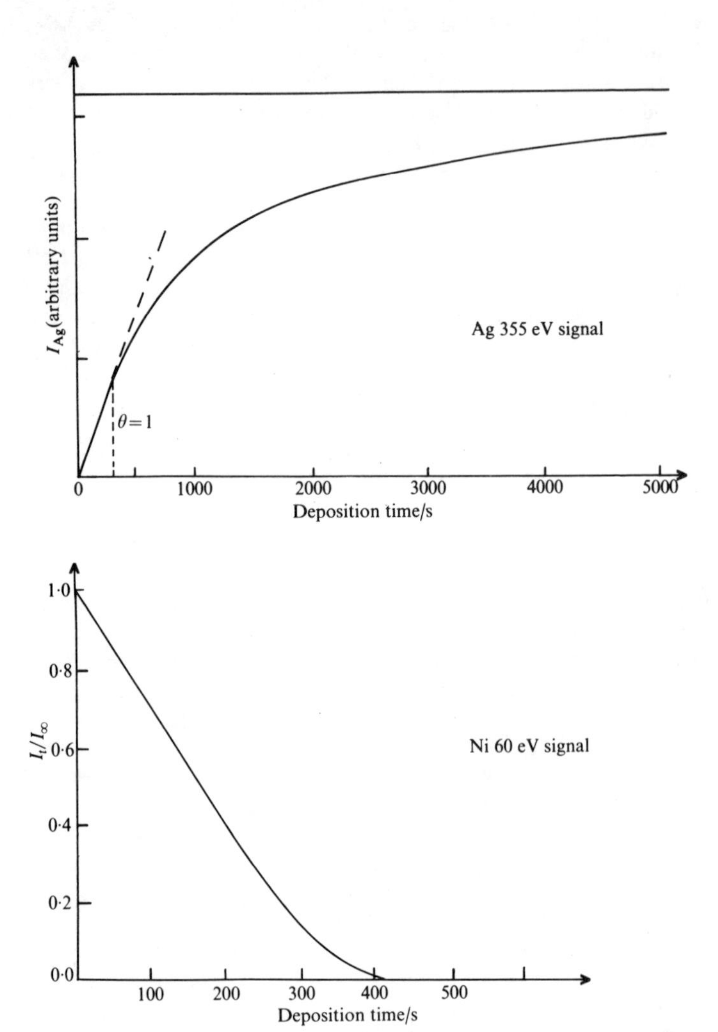

FIG. 2.11. The growth of a silver MNN Auger signal at 355 eV and decay of a nickel – MVV Auger signal at 60 eV as silver is deposited at constant rate upon Ni(100). The silver grows up monolayer by monolayer. The break point in the rising curve for silver occurs at a coverage of one monolayer because at greater coverages the outermost silver atoms shield the first monolayer. (After Jackson, Chambers, and Gallon 1973).

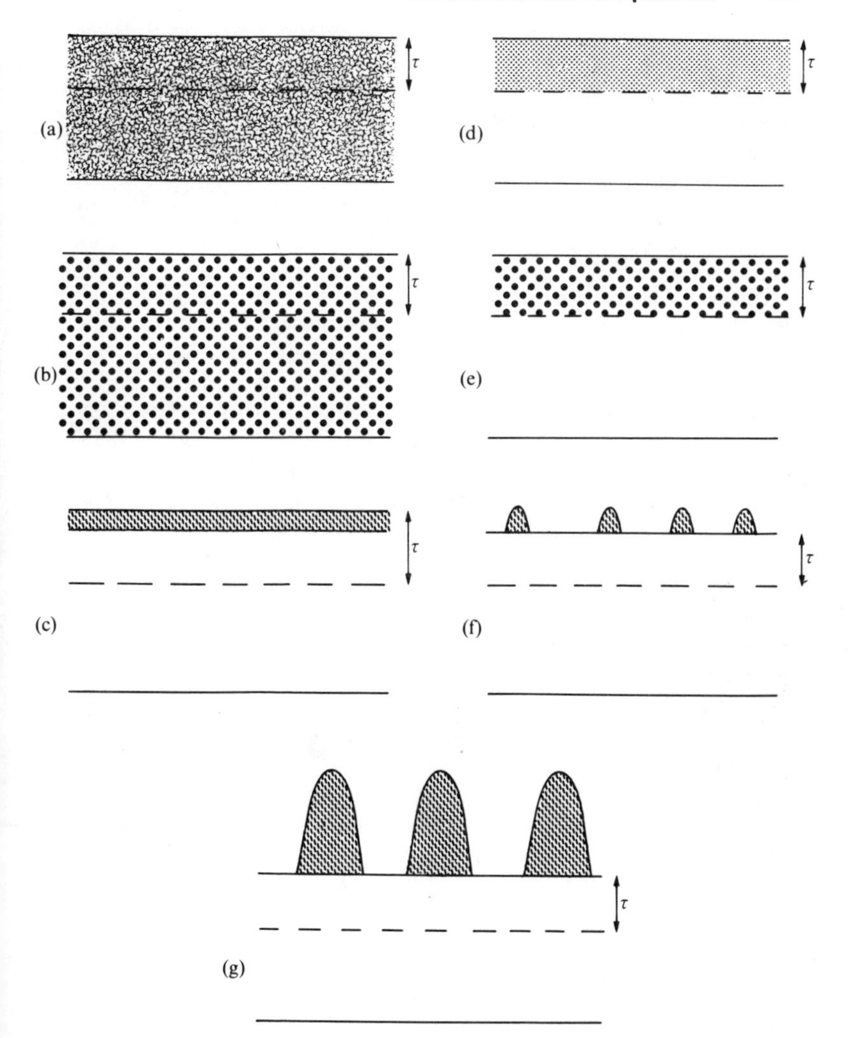

FIG. 2.12. Different arrangements of atoms of type A with substrates of type B which may occur in practice and which complicate the quantitative use of AES or XPS. (a) Uniform solution of atoms of A in matrix of B – e.g. an alloy. (b) Uniform distribution of clumps of atoms of A in matrix of B – precipitation. (c) Solutions of atoms of A only inside in the escape depth τ in B – surface segregation. (d) Clumps of atoms A inside the escape depth τ in B – surface precipitation. (e) Uniform thin layer of A upon B – a thin film. (f) Islands of atoms A upon B – e.g. some thin films in the initial stages of growth. (g) Tall needles (of height greater than τ) of A upon B – e.g. acicular growth of one substance upon another.

distributed through the whole sample. Examples of practical situations are indicated in Fig. 2.12, and it is easily seen that measurement of Auger peak height alone will not resolve the difference between, for example, Fig. 2.12(d) and Fig. 2.12(e). Other techniques have to be brought to bear in addition to electron spectroscopy to help resolve the difficulties of topography and distribution.

Other methods

There are many other methods of surface analysis (e.g. ionization-loss spectroscopy and appearance-potential spectroscopy) which will not be described here. Two particularly sensitive techniques should be mentioned, however.

1. *The atom probe.* By combining a field ion microscope (see Chapter 3) with an aperture and a mass spectrometer, Müller and Tsong (1969) were unable to detect and identify (by the ratio m/e) individual atoms. This is the ultimate in chemical sensitivity, but it is limited to rather special experimental geometries suitable for the field ionization of atoms.

2. *Secondary ion mass spectroscopy.* In this technique ions of a rare gas (Ar^+, Ne^+, He^+) are formed into a beam focused upon the surface of a solid. If the incident ions are sufficiently energetic (above say 500 eV) they interact in a nearly classical manner with the surface atoms knocking them into the vacuum with billiard-ball-like collisions. Some of the ejected particles are ionized and can be detected by a mass spectrometer. The high sensitivities of quadrupole mass spectrometers and electron multipliers can be used to detect these secondary ions and the incident ion flux kept so low that the removal rate of surface atoms is nearly negligible. Sensitivities of 10^{-6} of a monolayer are possible with this nearly non-destructive technique (Benninghoven and Loebach 1971).

The various techniques for characterizing surfaces are summarized at the end of Chapter 3.

3. Surface structure

Having determined what types of atoms are present upon a surface the next important problem is to discover their arrangements with respect to each other and with respect to the underlying atoms of the solid. There are two parts to this problem – the determination of the symmetry of the surface atomic arrangement and the determination of details of the atomic positions. In bulk investigations, the former is normally carried out by using simple observations of the diffraction pattern obtained when a beam of X-rays, neutrons, or electrons are scattered from a single crystal of the sample. These observations yield information about the symmetry of the repeating unit of the structure, the unit cell, and the size and shape of this cell. The latter involves measurements of the intensity of diffracted beams and the comparison of these intensities with those predicted from postulated models of the structure. If successful, the result is a complete description of where each kind of atom is situated within the unit cell.

Bulk techniques for structure analysis

X-ray diffraction is by far the most common method for studying bulk structures. X-rays are scattered by the charge distribution in and around atoms, and because this scattering is very weak they can penetrate materials very deeply and the bulk structure can be probed. By the same token the small atomic scattering cross-sections of atoms for X-rays result in a relative insensitivity to surface atoms. Using high-intensity X-ray sources and sensitive X-ray detectors, it has been possible to study the structure of metal films 10 nm thick but the difficulties of extending this work to smaller thicknesses are very considerable. Nevertheless, in spite of this lack of suitability of X-rays for surface structural studies, it is useful to examine in a simplified way the principal steps in a bulk X-ray structure determination of a simple solid.

A possible sequence of events for a structure determination is outlined in Fig. 3.1. Having obtained a single crystal of the sample and determined its composition (a combination that may be by no means trivial) a Laue diffraction pattern is obtained (e.g. Wormald 1973). Combining these observations with the Ewald sphere construction (Fig. 3.2) the unit cell of the structure can be determined. In finding the unit cell it is particularly useful to be aware of systematic absences. These arise whenever the direction of diffraction corresponds to destructive interference between the scattered waves. These absences can be found by using the *structure factor* F_{hkl} of the unit cell. The intensity I_{hkl} of a spot with Miller indices (hkl) is related to F_{hkl} through

$$I_{hkl} \propto |F_{hkl}|^2, \tag{3.1}$$

and, for cubic structures, F_{hkl} is given by

$$F_{hkl} = \sum_p f_p \exp 2\pi i(\mu h + \nu k + \omega l). \tag{3.2}$$

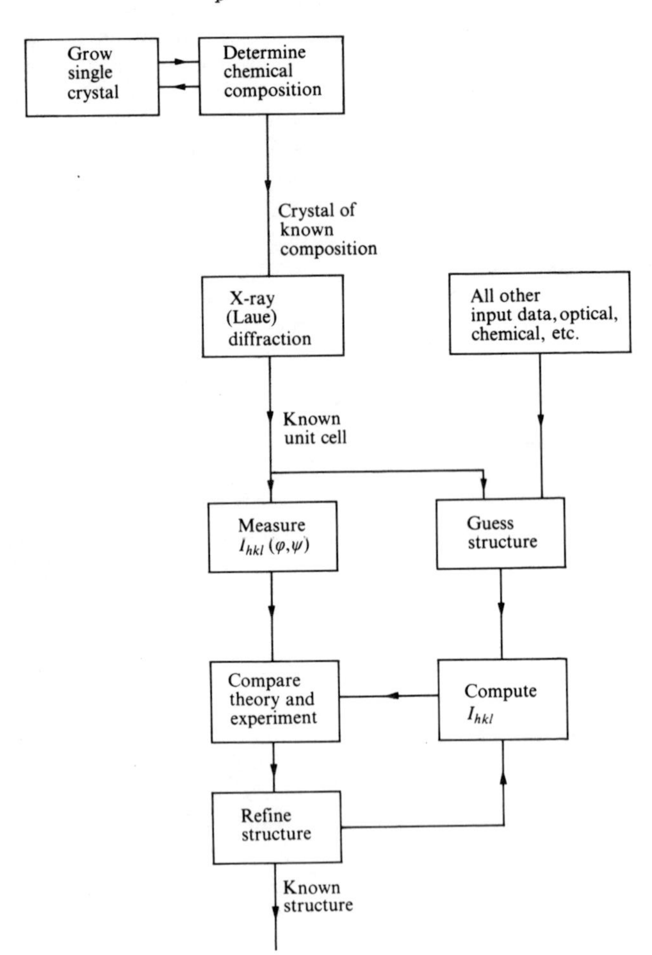

FIG. 3.1. A simplified sequence of steps for the determination of a structure.

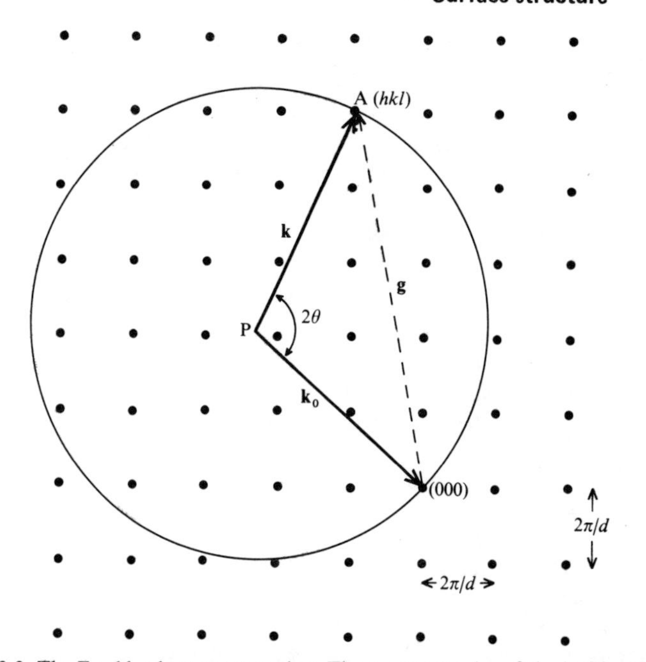

FIG. 3.2. The Ewald sphere construction. The wavevector $\mathbf{k}_0$ of the incident radiation is drawn on a scale diagram of the reciprocal lattice of the structure. The tip of $\mathbf{k}_0$ is at the origin (000) of reciprocal space. A sphere drawn with radius k_0 and with centre P is the Ewald sphere. If a point on the reciprocal lattice lies on the surface of the Ewald sphere, the condition for elastic scattering (the Bragg equation) is satisfied and diffraction will occur with the scattered beam having wavevector $\mathbf{k}'$. For monochromatic X-rays and arbitrary $\mathbf{k}_0$ it is clear that very few diffracted beams are excited (only beam PA in the case shown). The case shown is for a simple cubic lattice. 2θ is the scattering angle. $\mathbf{g}$ is the reciprocal lattice vector and (hkl) the Miller indices of the point on the Ewald sphere.

By using $|k| = 1/\lambda$, $|\mathbf{k}_0| = |\mathbf{k}'|$, and by dropping a perpendicular from P onto $\mathbf{g}$ it can be shown that the construction is equivalent to

$$n\lambda = 2d \sin \theta \qquad \text{(Bragg's law)}$$

or

$$\mathbf{k}' = \mathbf{k} + \mathbf{g} \ ,$$

where λ is the wavelength and $n = \sqrt{(h^2 + k^2 + l^2)}$.

In this equation the sum is over the atoms in the unit cell; f_p are the atomic scattering factors for each type of atom p in the structure; μ, ν, ω are the positions of the atoms in the unit cell expressed as fractions of the cell sides.

Some examples of simple systematic absences are:

1. in the f.c.c. system where there are no diffracted beams with (hkl) mixed even and odd,
2. in the body-centred cubic (b.c.c.) system only beams with $(h + k + l)$ even are found – all others are systematic absences;
3. in the sodium-chloride-type structure there are bright beams when (hkl) are all even, dim beams for (hkl) all odd, and systematic absences for all other (hkl).

These results can be demonstrated using eqn (3.2), and similar rules exist for all structures. It is possible to learn to recognize these features of diffraction patterns after practice.

By measuring the angles between diffracted beams and knowing the wavelength λ, the sides of the unit cell can be determined by using Bragg's law.

The difficult step in structure analysis comes after the identification and measurement of the unit cell when it is decided to try to find out where the atoms are within the unit cell. Fig. 3.1 indicates the kind of procedure that can be adopted to tackle this part of the problem. The difficulty arises from the fact that the experimental measurement must be of the intensity of the diffracted wave and this involves losing information about its phase. Because of this loss of information the diffraction pattern cannot be unravelled to reveal the structure directly, but other information, obtained in different experiments, has to be employed. For instance, chemical reactivity and spectroscopic observations can give information about bonding between the atoms present, and the natural occurrence of certain crystal faces suggests that these have low free energies and this observation may imply that particular combinations of atoms may be in these faces. All such guides are used together with the unit-cell determination to propose a model structure and make predictions, through equations like (3.2), of the intensities of diffracted beams as a function of the diffraction geometry (Fig. 3.3). These predictions are compared with a detailed set of measurements of the intensities I_{hkl} of a large set of beams (hkl) as functions of the angles ϕ and ψ explained in Fig. 3.3. This comparison enables more refined models of the structure to be made, and the loop is repeated until satisfactory agreement between theory and experiment is obtained.

The same procedure can be used to solve structures by observing the elastic scattering of neutrons in crystals. However, the scattering cross-sections of atoms for neutrons are lower than those for X-rays and, consequently, neutrons scattering is even less surface sensitive than X-ray scattering.

A form of incident wave which is much more strongly scattered by

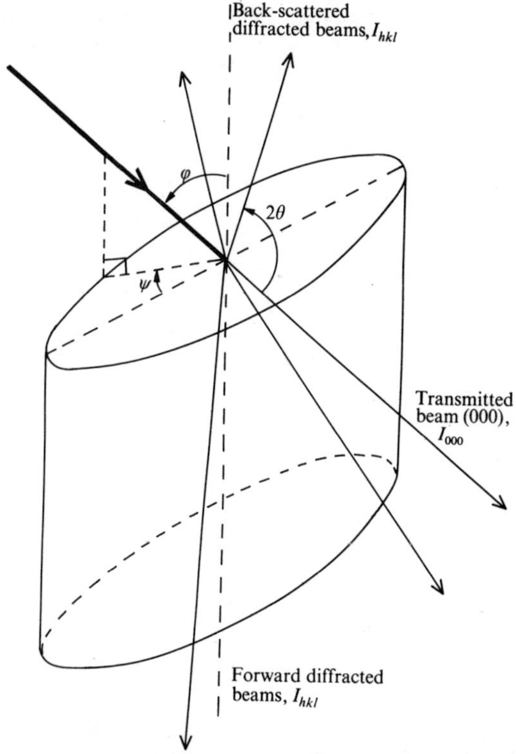

FIG. 3.3. Diffraction geometry from a cylindrical sample. ϕ is the angle of incidence measured from the normal to the surface upon which the radiation impinges. ψ is the azimuthal angle. It is measured from a reference direction in the surface (perhaps a low-index crystal direction) to the projection of the incident beam onto the surface. 2θ is the scattering angle.

atoms than either neutrons or X-rays is one of electrons. The atomic scattering probabilities for three processes are indicated as a function of their energies in Fig. 3.4. Electrons are usually at least 10^3 times more strongly scattered than X-rays and are particularly strongly scattered if their energies are below 1000 eV. It is this strong scattering cross-section that makes electrons suitable probes for surface structure studies.

Surface methods using electrons

The two important techniques for surface structure analysis using electrons are low-energy electron diffraction (LEED) and reflection high-energy electron diffraction (RHEED). Both methods can be used to determine the periodic two-dimensional arrangement of atoms at the surface – the unit mesh. A number of such arrangements are possible and it is useful to have some notation to describe this unit mesh before discussing the experimental methods.

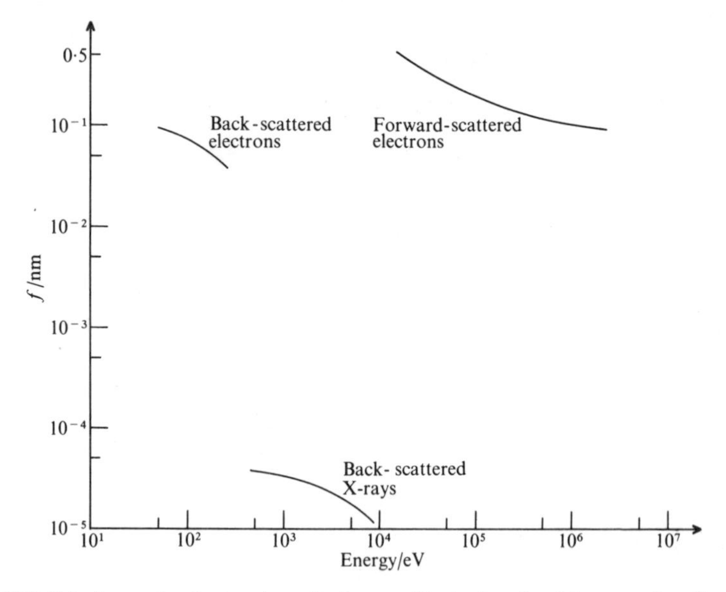

FIG. 3.4. Approximate atomic scattering amplitudes for aluminium as a function of energy of electrons and X-rays,

Notation for surface structures

Just as bulk, triperiodic structures can be divided into fourteen groups corresponding to their different Bravais lattices (Rosenberg 1974) so two-dimensional periodic structures can be grouped into five types of *surface nets.* These five nets describe all possible diperiodic surface structures. It is convenient to specify the surface itself by the Miller indices (hkl) of its normal. Then Fig. 3.5 shows the unit areas or *unit meshes* of the five possible nets. The complete net can be generated by translating the unit mesh parallel to the vectors $\mathbf{a}_{1s}$ and $\mathbf{a}_{2s}$ an integral number of times. The subscript s will be used for a surface mesh and the subscript b for a bulk exposed plane.

It is convenient to be able to relate the translations $\mathbf{a}_{1s}$ and $\mathbf{a}_{2s}$ of a surface net area A to the translations $\mathbf{a}_{1b}$ and $\mathbf{a}_{2b}$ of a second net of area B. This is required, for instance, if a reconstructed surface is to be

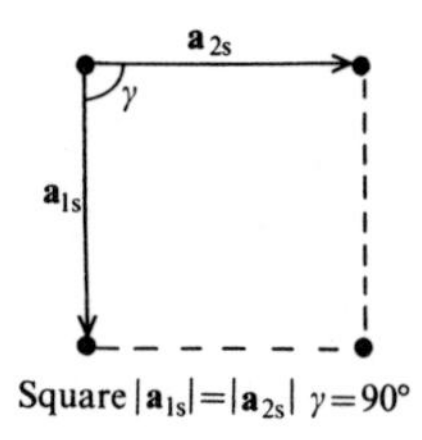

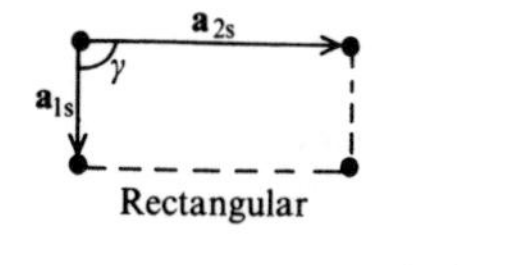

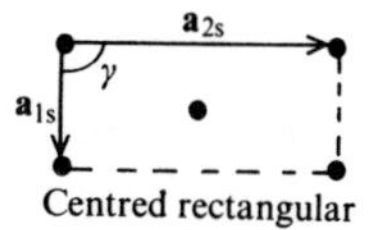

$|\mathbf{a}_{1s}| \neq |\mathbf{a}_{2s}|$ $\gamma = 90°$

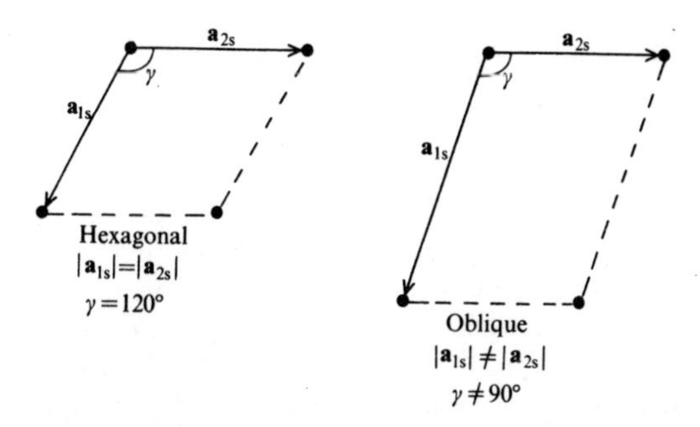

FIG. 3.5. Unit meshes of the five possible surface nets.

described in terms of the bulk exposed plane (Chapter 1) or if a monolayer of some surface deposit has a different unit mesh from that of the solid surface upon which it is adsorbed. This relationship is made most powerfully by using a transformation matrix **M** such that

$$\mathbf{a_s} = \mathbf{M}\,\mathbf{a_b}, \tag{3.3}$$

where

$$\mathbf{M} = \begin{bmatrix} m_{11} & m_{12} \\ m_{21} & m_{22} \end{bmatrix} \tag{3.4}$$

and thus

$$\left.\begin{aligned} \mathbf{a}_{1s} &= m_{11}\mathbf{a}_{1b} + m_{12}\mathbf{a}_{2b} \\ \mathbf{a}_{2s} &= m_{21}\mathbf{a}_{1b} + m_{22}\mathbf{a}_{2b} \end{aligned}\right\} \tag{3.5}$$

Since the areas of the two unit meshes are given by

$$A = |\mathbf{a}_{1s} \times \mathbf{a}_{2s}| \tag{3.6}$$

and

$$B = |\mathbf{a}_{1b} \times \mathbf{a}_{2b}|\,, \tag{3.7}$$

eqn (3.5) can be inserted into (3.6) to yield (using 3.7))

$$A = B \det \mathbf{M}. \tag{3.8}$$

The values of det **M** can be used to define the type of superposition which exists between the surface mesh $\mathbf{a}_s$ and the bulk mesh $\mathbf{a}_b$. This definition and the cases of simple, coincidence, and incoherent superpositions are described in Fig. 3.6. Although an example is not shown in Fig. 3.6 for an incoherent superposition cases of this type are found in practice, and they correspond to the phenomenon of epitaxial growth described in Chapter 6.

Although the matrix notation is powerful and general, many observed superpositions have been described by a simple notation devised by Wood (1964). Here the meshes $\mathbf{a}_s$ for the surface and $\mathbf{a}_b$ for the bulk are related by the ratios of the lengths of the translation vectors and by a rotation R expressed in degrees. Thus the meshes are related by an expression of the form $(\mathbf{a}_{1s}/\mathbf{a}_{1b} \times \mathbf{a}_{2s}/\mathbf{a}_{2b})R$. Two examples of this notation are illustrated in Fig. 3.7. If the deposit mesh is not rotated with respect to the substrate (or bulk) mesh then the rotation is simply dropped from the notation. Similarly a statement as to whether the mesh is primitive or centred (Fig. 3.5) should be included in the notation, but it is usual practice to drop any symbol for a primitive mesh and to insert a lower-case c before the statement of translation vector ratios for a centred mesh. This notation becomes clearer as examples are studied.

Diffraction from diperiodic structures

Because the periodicity along the surface normal is lost in a two-dimensional arrangement of atoms, the constructive interference of scattered waves cannot occur in this direction. This relaxation of the conditions for diffraction leads to the possibility of diffracted beams occurring at all energies and hence the fact that a diffraction pattern can be observed at all energies and in any geometry. This can be understood by using

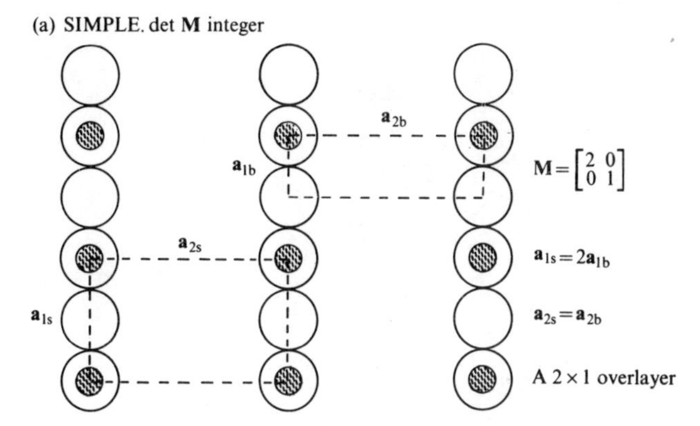

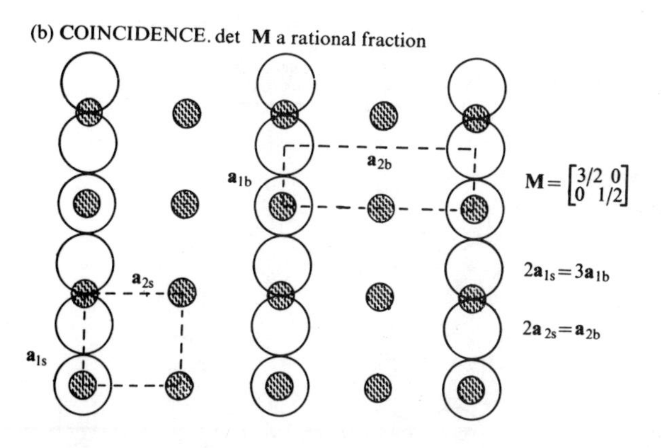

(c) INCOHERENT. det **M** an irrational fraction
$\mathbf{a}_s$ and $\mathbf{a}_b$ irrationally related

FIG. 3.6. Relationships between surface and bulk meshes. The simple and coincidence meshes are illustrated by the cases of deposit atoms (hatched circles) on the bulk exposed (110) plane of an f.c.c. material (open circles).

again the Ewald sphere construction in a reciprocal lattice diagram for a two-dimensional net of atoms. A rather loose view of this reciprocal-space diagram can be obtained by realizing that distances in reciprocal space are inversely proportional to distances in real space, and so if a triperiodic lattice is extended along one of its axes then reciprocal lattice points move closer together along this axis. In the extreme of this extension only one plane of atoms is left (the others being removed to infinity)

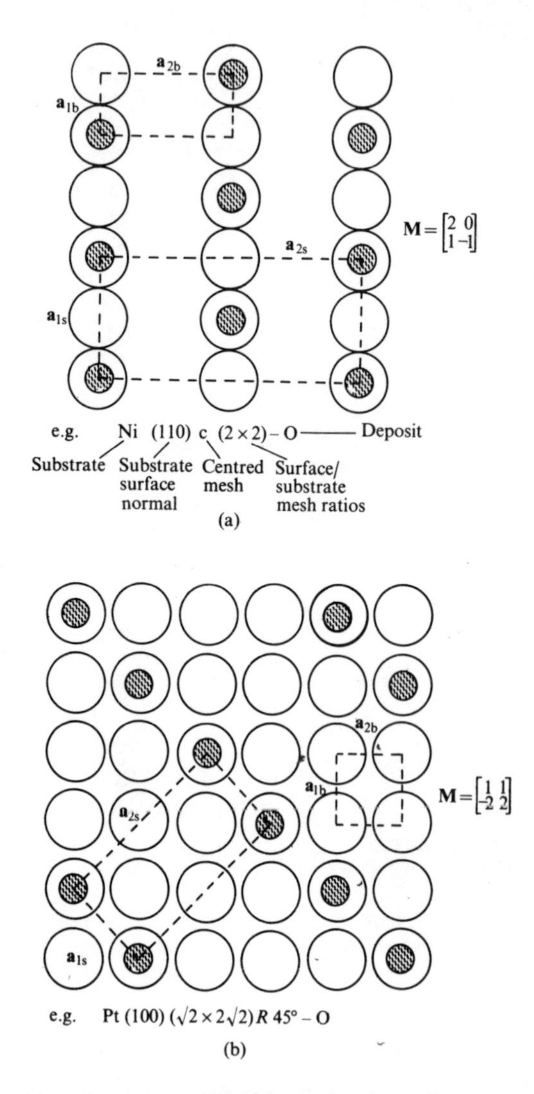

FIG. 3.7. Two notional examples of Wood's notation for surface structures compared with the matrix notation. In example (a) for a Ni(110) face exposed to oxygen the notation can be shortened slightly to Ni(110)c2–O because the deposit mesh is square.

and the reciprocal lattice has become a set of infinitely long rods normal to the plane of atoms.†

The spacings $\mathbf{a}_1^*$ and $\mathbf{a}_2^*$ of the unit mesh of the reciprocal lattice are related to the spacings $\mathbf{a}_1$ and $\mathbf{a}_2$ of the real lattice by the equation

$$\mathbf{a}_i^* \cdot \mathbf{a}_j = 2\pi\delta_{ij},$$
$$\delta_{ij} = 1, i = j$$
$$\delta_{ij} = 0, i \neq j. \quad (3.9)$$

2π is chosen on the right-hand side of eqn (3.9) (instead of unity, as conventionally used in X-ray diffraction) because it enables the wave vectors **k** to be drawn upon the reciprocal-lattice diagram and the Bragg equation (Fig. 3.2) to be employed directly. This is because the energy E of the incident electron beam is given by

$$E = (h^2/2m)k^2 \quad (3.10)$$

and

$$k = 2\pi/\lambda. \quad (3.11)$$

The Ewald sphere construction can then be applied to the diperiodic diffraction problem as shown in Fig. 3.8. Because of the loss of periodicity in one dimension, the reciprocal-lattice rods can be labelled with only two Miller indices h and k, and a general reciprocal lattice vector $\mathbf{g}_{hk}$ lies in the plane of the surface and is given by

$$\mathbf{g}_{hk} = h\mathbf{a}_1^* + k\mathbf{a}_2^*. \quad (3.12)$$

Diffraction occurs everywhere the Ewald sphere cuts a reciprocal-lattice rod and the diffracted beam can be labelled with the Miller indices (hk) of the rod causing it. Again because of the loss of periodicity in one dimension, the vector diffraction equation becomes

$$\mathbf{k}_{\parallel}' = \mathbf{k}_{0\parallel} + \mathbf{g}_{hk} \quad (3.13a)$$
$$|\mathbf{k}'| = |\mathbf{k}_0|, \quad (3.13b)$$

where the subscript $\parallel$ means the component of a vector parallel to the surface. Eqn (3.13a) amounts to conservation of the component of

†A more rigorous method of deriving this result is via the use of the fact that the reciprocal lattice is the Fourier transform of the real lattice (Woolfson 1971). As the Fourier transform of a plane is an infinite rod normal to the plane so the Fourier transform of a coplanar set of points is a set of infinite rods normal to the plane.

momentum parallel to the surface and its validity can be seen in Fig. 3.8. Eqn (3.13b) amounts to the conservation of energy used here because it is an elastic scattering process under consideration.

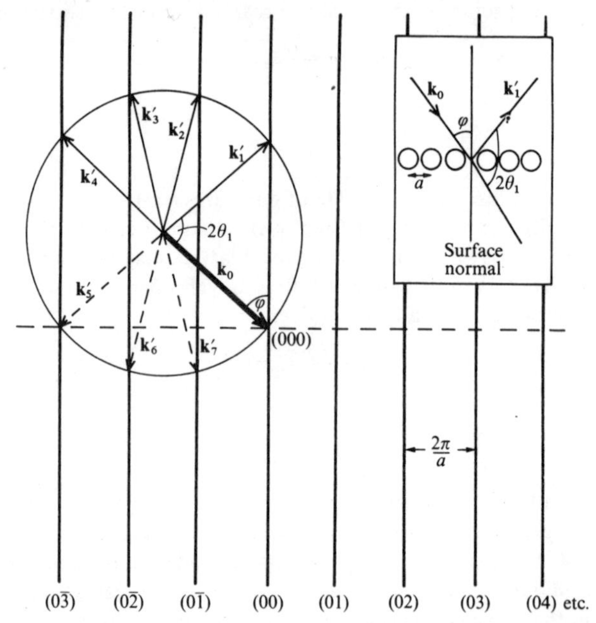

FIG. 3.8. The Ewald sphere construction (e.g. Wormald 1973) applied to diffraction from a square net of atoms of side a. In the case drawn seven elastically scattered diffracted beams can be generated by incident wavevector $\mathbf{k}_0$ arriving at an angle of incidence ϕ to the surface normal. Four beams are shown back-scattered from the surface; three beams are shown transmitted into the solid. More than this number of beams will occur because only that part of reciprocal space in the plane of the paper can be drawn. *Inset*. Real space diagram of the specular beam $\mathbf{k}_1{}'$.

Reflection high-energy electron diffraction (RHEED)

If a beam of high-energy electrons is incident upon a flat surface in grazing incidence (Fig. 3.9) the diffraction pattern formed will be characteristic of the surface atomic arrangement because the component of the incident electron momentum normal to the surface is very small and thus the penetration of the electron beam will be small.

At high energies the wavelength of the electron is small and the radius of the Ewald sphere is large compared to typical reciprocal-lattice vectors. Thus, at 100 keV, $\lambda = 0{\cdot}0037$ nm and $\mathbf{k}_0 = 1700\ \mathrm{nm}^{-1}$, whereas $2\pi/a$ might typically be 20 nm. Compared to the reciprocal lattice the Ewald sphere is very large and will cut the (00) rod almost along its length, as indicated in Fig. 3.9(b). Only those rods in the plane normal to the paper and containing the (00) rod will contribute beams to the diffraction pattern. This RHEED pattern will therefore consist of long streaks normal to the shadow edge of the sample and spaced by a distance *d*.

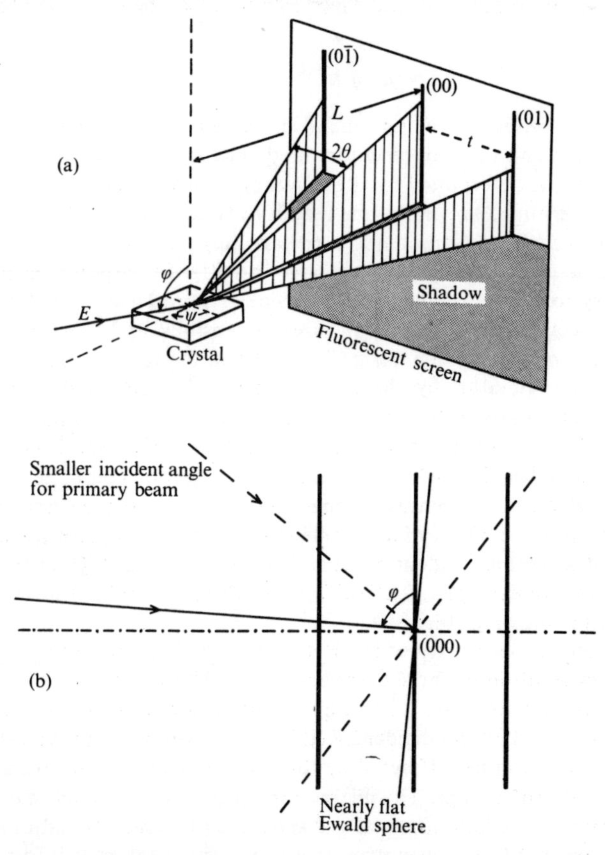

FIG. 3.9. The RHEED method. (a) Experimental goemetry. A fine parallel beam of electrons is incident near $\phi = 90^\circ$ upon a flat single-crystal surface. (b) Ewald sphere construction for RHEED.

If the separation between the fluorescent screen and the sample of some cubic crystal is L (the camera length) and the distance between streaks on the screen is t then

$$t = L \tan 2\theta \qquad \text{by geometry}$$

and

$$\lambda = 2d \sin\theta = \frac{2a}{(h^2 + k^2)^{\frac{1}{2}}} \sin\theta \qquad \text{by Bragg's law,}$$

for a sample with a square lattice of side a. As $\lambda \ll a$ in RHEED, then the values of 2θ are small and these equations can be simplified to yield

$$a = (h^2 + k^2)^{\frac{1}{2}} \; \lambda L/t\,. \tag{3.14}$$

All the parameters on the right-hand side of eqn (3.14) can be deduced or measured and so a can be determined. The accuracy with which this can be done is determined by the accuracy with which t can be measured. This is determined largely by the length L. High-accuracy RHEED experiments are usually constructed so as to have long camera lengths for this reason.

Because the Ewald sphere is so large in a RHEED experiment it is necessary to change the diffraction geometry in order to find the arrangement of reciprocal-lattice rods in three dimensions, and thus to define the unit mesh. As can be seen using Fig. 3.9(b) all the reciprocal-lattice rods can be explored either by changing the angle ϕ by rocking the sample about an axis in its surface or by rotating the sample about its surface normal. The latter is to be preferred if constant surface sensitivity is to be maintained as the rocking experiment changes the component of the incident electron momentum normal to the surface. If the azimuthal angle (ψ in Fig. 3.3) is varied, simple streak diffraction patterns can be observed when the incident beam is along directions of high crystal symmetry. An example of RHEED patterns from a Si(111) surface in two azimuths is shown in Fig. 3.10.

The surface sensitivity of the RHEED method is affected by two factors in addition to those described above. The simplest of these two is the surface roughness. If small bumps or needles of material stick out of the surface then the incident RHEED beam will pass through them. In doing so it will be diffracted by the three-dimensional atomic arrangement within the bumps. The diffraction process is then much more like diffraction in the bulk material and spots are observed instead of streaks. On the one hand, this phenomenon is useful in that it gives information about the surface topography but, on the other, it is difficult in practice to obtain surfaces which are sufficiently flat to give clear streaked diffraction patterns like those in Fig. 3.10.

The second factor is associated with uncertainties in the wavevector of the incident electrons due to the finite convergence and finite energy spread in electron beams coming from real sources. These uncertainties reveal themselves as a region of finite extent over which the electrons can be regarded as being in phase with each other. This region is called

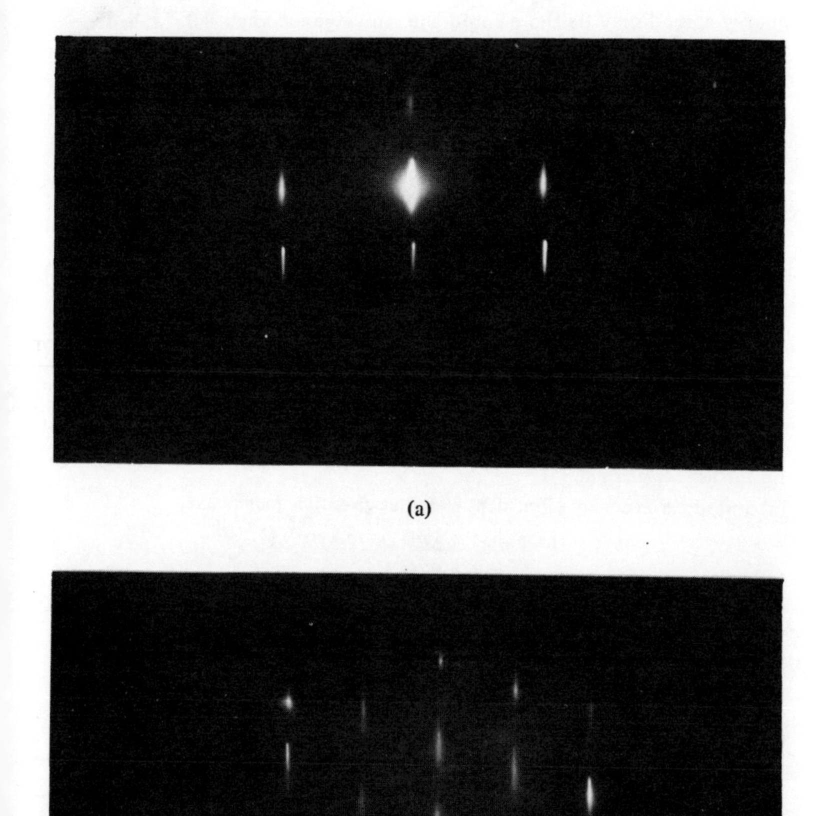

(a)

(b)

FIG. 3.10. RHEED patterns obtained at 100 keV from the (111) surface of silicon. (a) $[\bar{2}11]$ azimuth; (b) $[\bar{1}01]$ azimuth. The streaking indicates that the surface is flat.

the *coherence zone* and an expression for its size is derived in the caption to Fig. 3.11. In a typical RHEED experiment at 100 keV, the energy spread may be 0·5 eV and the convergence angle 10^{-5} rad. In these circumstances, time incoherence is negligible and the spatial incoherence zone is about 200 nm in diameter. Ordered regions on the surface very much smaller than this will broaden the diffracted beam to the

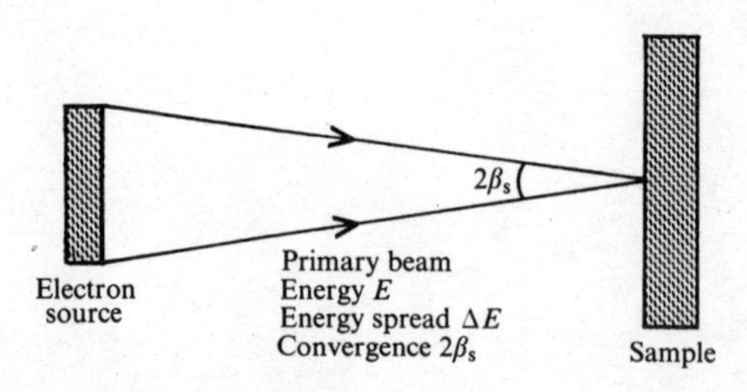

FIG. 3.11. The coherence zone diameter. See, for instance, Heidenreich (1964) for further explanation.

(a) Spread ΔE in energy of incident electrons gives *time incoherence.*

$$E = (h^2/2m)k^2\ , \quad \Delta E = (h^2/2m)2k\ \Delta k\ ,$$

therefore

$$\Delta k^{\mathrm{t}} = k\Delta E/2E\ .$$

Resolved parallel to surface, for small β_s,

$$\Delta k^{\mathrm{t}}_{\parallel} \simeq (k\Delta E/E)\beta_s\ .$$

(b) Spread in arrival angle over $2\beta_s$ gives *spatial incoherence,*

$$\Delta k^{\mathrm{s}}_{\parallel} \simeq k 2\beta_s\ .$$

(c) Combine uncertainties in quadrature and define *coherence zone diameter* ΔX,

$$\Delta X \Delta k_{\parallel} = 2\pi$$

$$\Delta k_{\parallel} = (\Delta k^{\mathrm{t}\,2}_{\parallel} + \Delta k^{\mathrm{s}\,2}_{\parallel})^{\frac{1}{2}}\ .$$

$$\boxed{\Delta X \simeq \frac{\lambda}{2\beta_s\left\{1 + (\Delta E/2E)^2\right\}^{\frac{1}{2}}}}$$

point where it will not be observable. If they are large compared to 200 nm then sharp well-defined diffraction patterns will be obtained. A RHEED pattern of half a monolayer of oxygen on Cu(110) is shown in Fig. 3.12.

The application of RHEED to full surface-structure analysis will be discussed briefly on p. 57.

Low-energy electron diffraction (LEED)

The high atomic scattering cross-sections for electrons with energies less than 1000 eV (Fig. 3.4) suggest that low-energy electron diffraction should be extremely sensitive to surface atomic arrangements. This has been known for a very long time – indeed Davisson and Germer used LEED in 1927 to demonstrate the wave nature of the electron, However, it is only since the development of UHV technology that LEED has been widely studied and used.

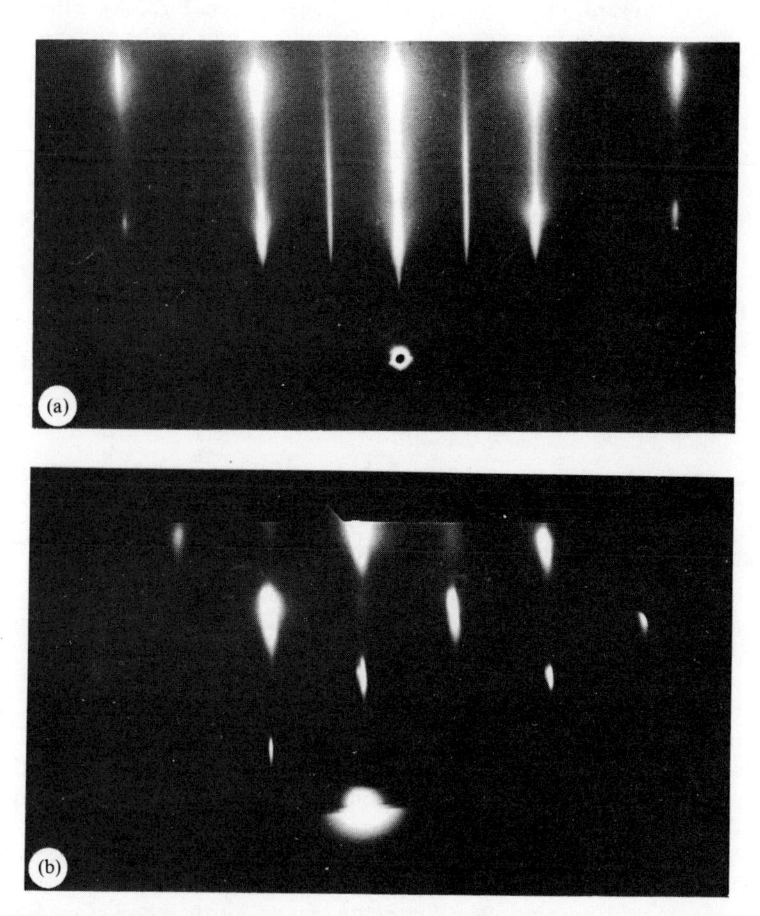

FIG. 3.12. RHEED at 100 keV from Cu(110)c(2 × 2)–O. The coverage corresponds to large flat islands of oxygen at a level of about half a monolayer. (By courtesy of F. Grønlund and P. E. Højlund Nielson.)

The Ewald sphere construction of Fig. 3.8 can be used to describe the electron beams that will be diffracted from a single-crystal surface. If only the top monolayer of atoms are responsible for the scattering then a beam with wavevector $\mathbf{k}_0$ incident at an angle ϕ will produce back-scattered beams $\mathbf{k}_1'$, $\mathbf{k}_2'$, $\mathbf{k}_3'$, and $\mathbf{k}_4'$ returning to the vacuum and forward-scattered beams $\mathbf{k}_5'$, $\mathbf{k}_6'$, and $\mathbf{k}_7'$ going on into the solid. By varying the primary energy (and so varying the radius $|\mathbf{k}_0|$ in Fig. 3.8) the number and directions of the scattered beams will vary. The single-crystal surface will thus produce a spot diffraction pattern which will contract towards the specularly reflected beam ($\mathbf{k}_1'$ in Fig. 3.8) as the primary energy is increased. If this diffraction pattern can be observed and the angles 2θ between the scattered beams and the incident beam measured then the surface unit mesh can be determined using Fig. 3.8 (which is drawn for a square mesh). Geometrical consideration (using the same arguments as those under Fig. 3.2) of this figure gives

$$a = \frac{(h^2 + k^2)^{\frac{1}{2}}\,\lambda}{2 \sin\theta \cos(\theta + \phi)} \tag{3.15}$$

for any particular reciprocal-lattice rod (hk). In the simpler case of normal incidence ($\phi = 0$) eqn (3.15) becomes

$$a = \frac{(h^2 + k^2)^{\frac{1}{2}}\,\lambda}{\sin 2\theta}, \tag{3.16}$$

and the slope of a plot $\sin 2\theta$ versus λ ($\lambda = \sqrt{150/10V}$ nm) can be used to find the mesh side a. When $\theta = 0$, Fig. 3.8 shows how a simple symmetrical LEED pattern will be observed for normally incident electrons.

A LEED pattern can be displayed by using the arrangement shown in Fig. 3.13, which is the same electron optical arrangement used for Auger electron spectroscopy (Fig. 2.3(b)). A beam with small convergence ($2\beta_s \sim 0{\cdot}01$ rad, Fig. 3.11) and variable energy is diffracted from the crystal surface and back-scattered beams move through field-free space between the crystal and grid G1. Between G1 and the screen they are radially accelerated so that they are energetic enough to excite fluorescence in the screen S and the spots of light so created viewed or photographed through the window. The grids G2 and G3 are provided so that those electrons that have been scattered inelastically in the sample can be rejected and mainly elastically scattered electrons reach the screen. The potential V in Fig. 3.13 would be adjusted so as to reject these inelastic electrons since they contribute only to a diffuse background in the diffraction pattern. Two grids are used to help reduce the effects of penetration of the high positive potential on the screen through the open mesh of the grids. This potential penetration simply reduces the quality of the arrangement as an electron energy analyser.

The means of holding the sample at the centre of the grid system is the subject of considerable experimental ingenuity. If the diffraction geometry is to be varied then both the angle of incidence ϕ and the azimuthal angle ψ need to be adjustable; exploration of different parts of the

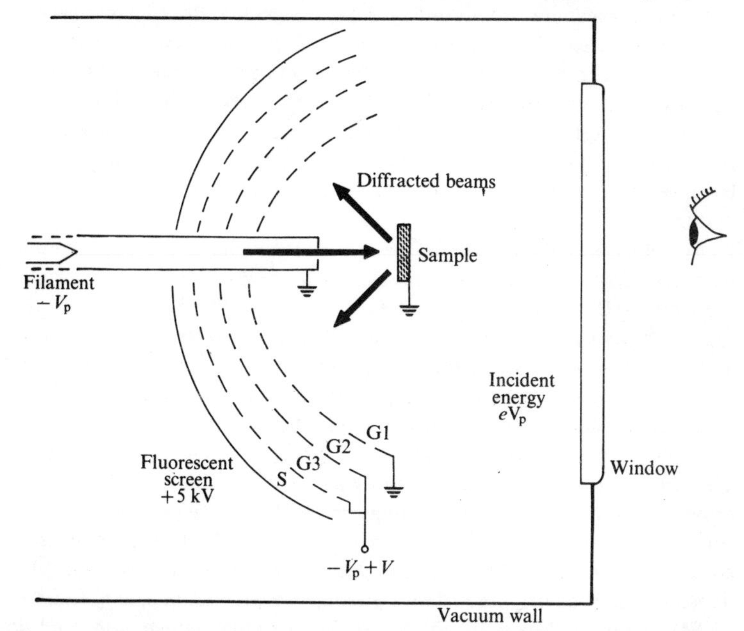

FIG. 3.13. LEED optics for displaying diffraction patterns. The screen S and the grids G1, G2, and G3 are spherical sections with a common centre at the point where the incident electron beam strikes the sample surface. The incident beam is usually focused to about 1 mm diameter at the sample, has a maximum current of about 2 μA, and an energy between 10 eV and 1000 eV. With different electrical connections the same electron optics can be used as a retarding potential analyser for electron spectrometry (Chapter 2).

surface requires two translations and adjustment of the surface to be at the grid centre requires a third translation; a study of diffraction intensities as a function of temperature (Chapter 5) may require both cooling and heating; cleaning up the sample surface may require heating to near the melting point of the sample or cleavage to produce a natural crystal face *in situ.* At the same time the observer's view of the diffraction pattern is blocked by any experimental arrangement around the sample, and consequently any of these facilities must be provided in the most unobtrusive way possible.

The symmetry of the atomic arrangement at a surface and the sides of the unit mesh (eqn (3.16)) can be obtained immediately in a LEED experiment without varying the diffraction geometry, as is necessary in RHEED. This is an important use of LEED because the progress of an experiment which changes the surface in some way can be quickly and conveniently followed. Some examples of LEED patterns from three types of surfaces are shown in Fig. 3.14 alongside indexed sketches of the same patterns. Fig. 3.14(a) is typical of most low-index metal surfaces after cleaning in that it shows the diffraction pattern expected from the bulk exposed plane of atoms. The clean surfaces of semiconducting materials are different to those of metals because they often give diffraction patterns indicating reconstruction (e.g. Fig. 3.14(c)). This is probably because of the directional character of the covalent bonds in these materials.†

In order to establish the surface sensitivity of AES, advantage could be taken of the fact that some metals grow upon others monolayer by monolayer (Chapter 2). The same fact can be used to examine the surface sensitivity of LEED. The systems Au(100)–Ag, Cu(100)–Ag, W(100)–Cu, and Cu(100)–Au all show this monolayer-growth process and the LEED features of the substrate are no longer detectable after one–six monolayers of the deposit and primary energies below 125 eV. The precise sensitivity depends upon the particular substrate–deposit combination. This kind of result has to be treated with care if the deposit does not grow with a monolayer habit. Using Fig. 3.11 and the parameters of typical LEED systems ($2\beta_s \sim 0{\cdot}01$ rad $\Delta E = 0{\cdot}5$ eV) then the coherence zone diameter is about 10 nm at a primary energy of 100 eV. Should a deposit consist of oriented islands greater than 10 nm in size then a LEED pattern will be seen which is difficult to distinguish from that due to a continuous oriented deposit.

By combining LEED and RHEED in the same apparatus the different diffraction geometries and coherence zone diameters can be used to draw conclusions about the surface topography and the extent of the ordered coverage as well as the size and symmetry of the surface mesh.

The high surface sensitivity of LEED arises from the large scattering cross-sections of atoms for low-energy electrons. This very property which makes LEED so useful for surface information also results in difficulties in the theoretical interpretation. Because the cross-section is high for elastic scattering at a single atom, it is possible for beams to be scattered several times and still emerge from the surface with measurable intensity (Fig. 3.15). The theory of X-ray diffraction, upon which rest the

†For further discussion of LEED patterns and their explanation see the review papers by Estrup and MacRae (1971) and Prutton (1971).

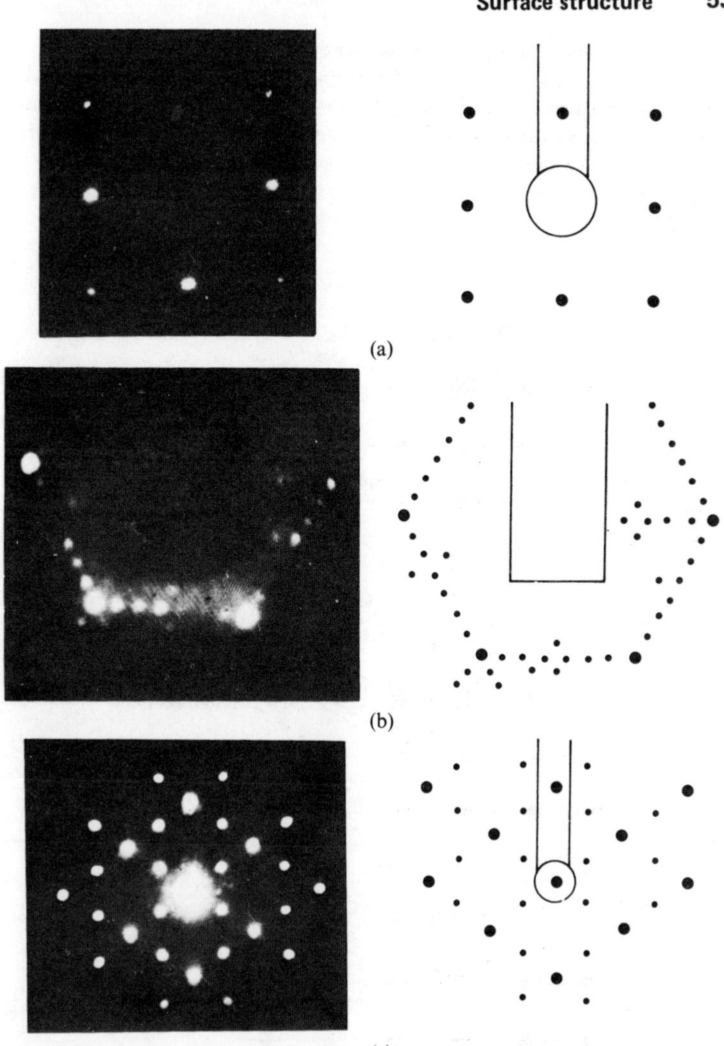

FIG. 3.14. Some typical LEED patterns. (a) The clean Cu(100) surface. Only spots from the bulk exposed plane are present. Primary energy 150 eV. (By courtesy of Dr R. J. Reid, New University of Ulster.) (b) The clean Si(111) surface. Extra spots are present between the six (10) features (bright spots) from the bulk exposed plane. These extra features correspond to a surface mesh parallel to the substrate mesh but with 7 times the length of its sides. Because of these $\frac{1}{7}$th order spots this pattern is called Si(111)(7 x 7) or Si(111)7. Primary energy 42 eV. (c) The W(110)-(2 x 1)–O LEED pattern due to oxygen adsorbed on W(110). Primary energy 53 eV. (By courtesy Dr J. W. May, Eastman–Kodak Laboratories, New York.)

methods described on pp. 33–6, is a *kinematical theory* in which the probability of such multiple scattering effects is treated as negligible. If multiple scattering occurs then all the waves scattered into a particular direction in many different scattering sequences must be added up with due regard for their correct amplitudes and phases. Such a treatment is referred to as a *dynamical theory* and it is essential in the description of LEED.

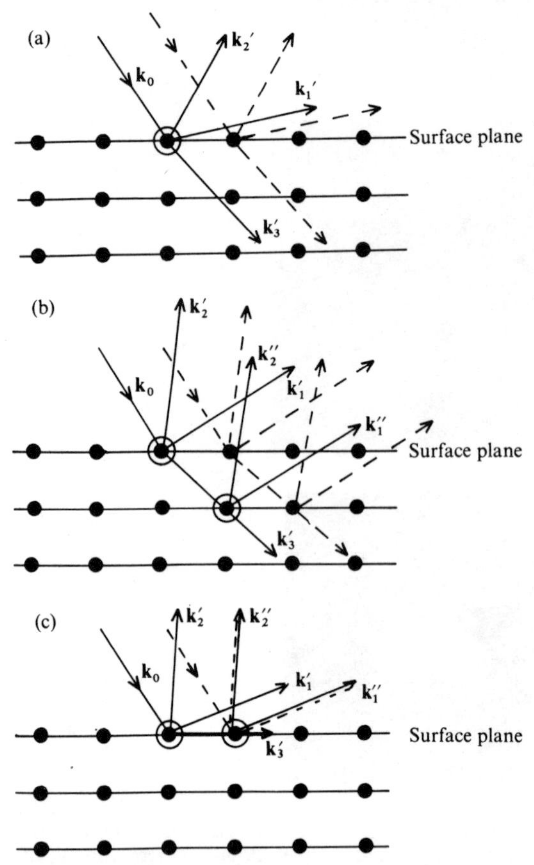

FIG. 3.15. The scattering of waves from a solid surface. The filled-in circles indicate the atomic positions and the open circles are added when a scattering event occurs. (a) Single scattering. Kinematical theory would describe such an event. Diffraction occurs when the 'dashed' wave adds up in phase with the 'solid' wave. (b) Double scattering. The forward-scattered beam $\mathbf{k}_3'$ is scattered into two beams $\mathbf{k}_1''$ and $\mathbf{k}_2''$ adding to beams $\mathbf{k}_1'$ and $\mathbf{k}_2'$ generated at the first scattering event. (c) A case of double scattering involving a surface wave $\mathbf{k}_3'$.

The occurrence of multiple scattering can be demonstrated in practice by two kinds of experiments. In the first the intensity of a particular diffraction spot is measured as a function of the primary energy. Such a plot is usually referred to as an $I(V)$ plot, and an example is shown in Fig. 3.16(a) for MgO(100). If normal three-dimensional diffraction had

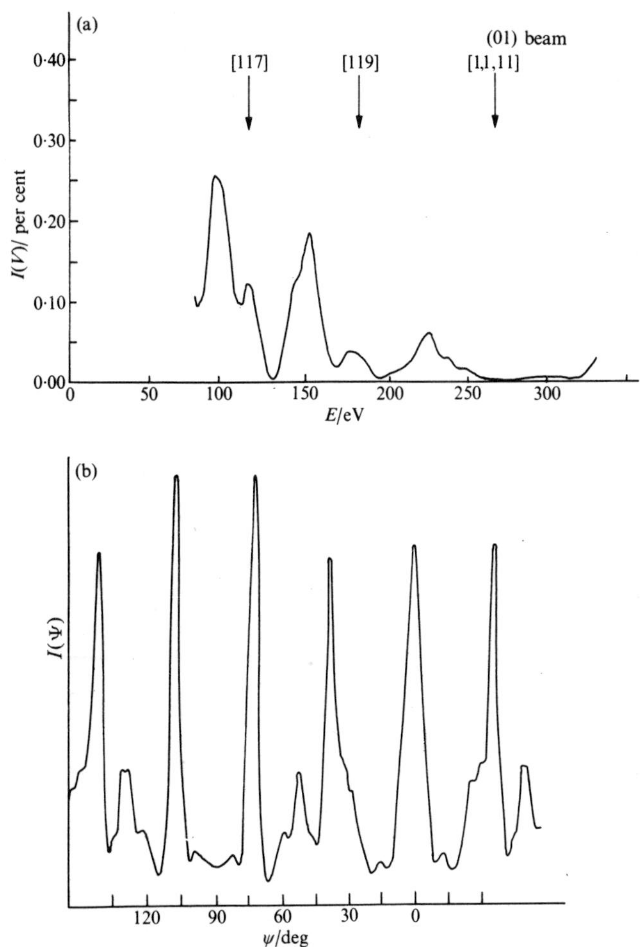

FIG. 3.16. (a) An $I(V)$ plot from Mg(100) in normal incidence. The energies of the Bragg peaks are marked and indexed. The shift between the large observed peaks and the calculated Bragg peak positions is due to the 'inner potential' in the crystal. (b) A rotation diagram I (ψ) from W(110). The primary energy is 595 eV and the diffraction geometry is set so that the (550) beam is being measured. (After Gervais, Stern, and Menes (1968).)

been occurring then a maximum intensity should occur whenever the incident wavelength had the correct value for Bragg's equation (Fig. 3.2) to be satisfied. The so-called 'Bragg peaks' are indicated in Fig. 3.16(a). It can be seen that extra peaks occur in the $I(V)$ plot. These are often called *secondary peaks* and are due to the multiple scattering. In the second experiment the primary energy is held constant and the crystal is rotated about its surface normal while the intensity of the (00) beam is measured. The graph of intensity I versus azimuthal angle ψ is called a *Renninger plot* or *rotation diagram,* and an example is shown in Fig. 3.16(b) for W(110). Looking at Fig. 3.8 it can be seen that such a rotation does not change the angle of incidence ϕ and a kinematical theory would suggest that $I(\psi)$ should be independent of ψ. However, if multiple scattering occurs then every time the value of ψ is such that strong diffraction can occur in some direction other than $\mathbf{k}_1'$, intensity must be lost from the beam $\mathbf{k}_1'$. Thus, the rotation diagram has minima everywhere the diffraction geometry is correct for strong scattering in some other direction from that being observed.

The theory of LEED and RHEED

Theoretical studies of LEED and RHEED usually have one of two objectives. They may be attempts to explain the measured intensities in $I(V)$ or $I(\psi)$ plots in LEED or in $I(\phi)$ plots in RHEED in terms of the properties of the solid – the kinds and positions of atoms present and the potential in which they are situated. Alternatively, they may be attempts to determine the atomic structure from the measured intensities of spots in patterns like those shown in Fig. 3.14. Of course, both objectives have a great deal in common, but the motivations, and therefore the parameters which are treated as adjustable, are very different.

Although measurements of the geometry of the diffraction pattern enable conclusions to be drawn about the size and symmetry of the unit mesh, it is only possible to discover where each kind of atom is situated within the mesh and where it is placed above the next layer of atoms by measuring the intensity of diffraction features as the diffraction parameters are varied. The procedure for structure analysis is thus very similar to that described in Fig. 3.1 except that the difficulty of computing I_{hk} and refining the guessed structure after comparison with the experimental data is very much greater than it is in the X-ray counterpart.

The theoretical situation in LEED is eased slightly by the fact that the mean free path before inelastic scattering is quite small (a fraction of a nanometre in this energy range (see Fig. 2.7, p. 23). This means that there is a restriction on the number of elastic scattering events that can occur before inelastic scattering spoils the coherence of the beam. This can be seen from Fig. 3.15, which indicates how the path length in the crystal of the elastically scattered beams increases as the number of scattering events increases. Thus a multiple-scattering theory in LEED does not require too many beams to obtain reasonable agreement with experiment. Nevertheless the theory of LEED is still very complex (Pendry 1974).

In RHEED the experimental inconvenience of varying the primary energy means that intensity data is usually in the form of rocking curves ($I(\phi)$) or rotation diagrams ($I(\psi)$). Although the scattering cross-sections are smaller in RHEED than in LEED and therefore multiple scattering might be expected to be less important, the mean free path before inelastic scattering is much longer. This has the effect that more scattered beams must be included for an accurate description of the intensity. The nett effect of smaller scattering cross-sections and longer mean free paths is to lead a similar theoretical difficulty.

Field ion microscopy (FIM)

A technique with the ability to detect directly the positions of atoms upon a surface was invented by Müller (1951) and is called 'field ion microscopy'. It is reviewed in Müller (1965). A diagram of a field ion microscope is given in Fig. 3.17. The specimen is prepared in the form of a sharp tip to which a positive potential is applied so that a field of the order of 5×10^8 V cm^{-1} is present at the tip surface. Molecules of the imaging gas (usually helium or neon at a pressure of $1-3 \times 10^{-3}$ Torr) move towards the tip and collide with it. After many collisions (Fig. 3.18) they are slowed down and loose an electron to the tip by quantum-mechanical tunnelling. When this happens the resultant positive gas ion is accelerated away from the tip in the large electrical field and strikes the fluorescent screen causing a spot of light to be created.

According to one simple model the ionization of the imaging gas atom is most likely to occur where the local electrical field is high, i.e. where the radius of curvature of the tip is highest. Thus a protruding atom is more likely to cause ionization of the imaging gas than a flat plane of atoms. The spatial resolution of the technique depends upon the component of the velocity of the imaging ions tangential to the tip surface. If the tip is cooled to reduce this component then resolutions of about 0·25 nm can be achieved. Therefore spots of light on the fluorescent screen correspond to the positions of individual atoms on the tip.

The material of the tip has to be a single crystal stable under the effect of high electrical field necessary to obtain ionization and has to be prepared so that needle tips having a radius of curvature well below 100 nm are possible. Tips of the elements W, Re, Ir, Pt, Mo, Ta, Nb, and Rh are all stable in the fields required to ionize helium, and the technique has been extended to tips of Zr, V, Pd, Ti, Fe, and Ni and some of their alloys with modifications to the imaging gas and the use of modern image-intensifiers to help observation of the weak picture on the fluorescent screen. The tip is prepared by etching in the laboratory and then

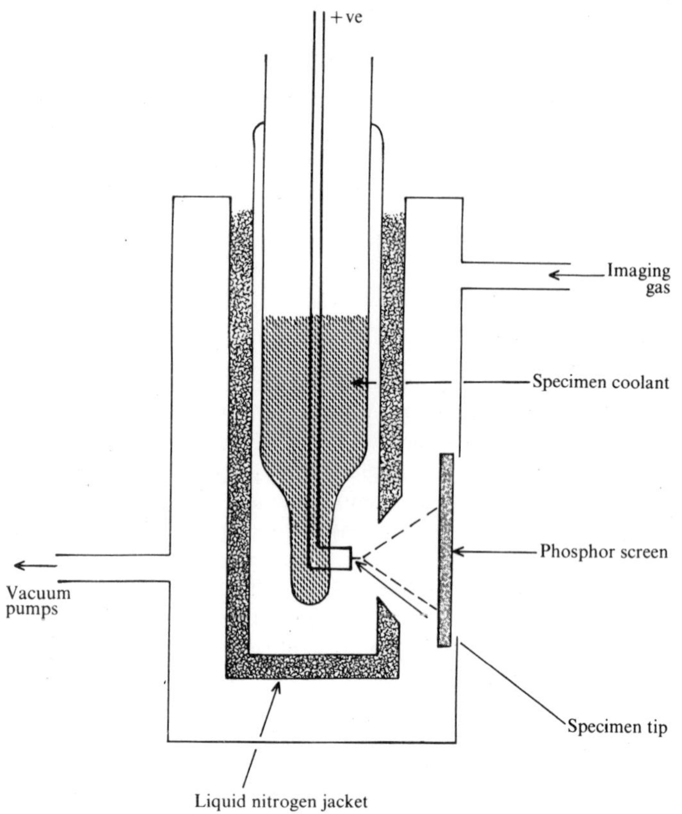

FIG. 3.17. A simple field ion microscope arrangement. The specimen tip is mounted on a wire which can be heated by passing a current through it.

formed *in situ* in the microscope by raising the applied electric field until surface atoms are removed as ions. This field evaporation process is self-regulating in that the field is greatest at sharp edges and aspherities and so these are removed first.

Fig. 3.19 shows a field ion micrograph of a platinum tip of 150 nm radius imaged with helium. The spots correspond to the protruding

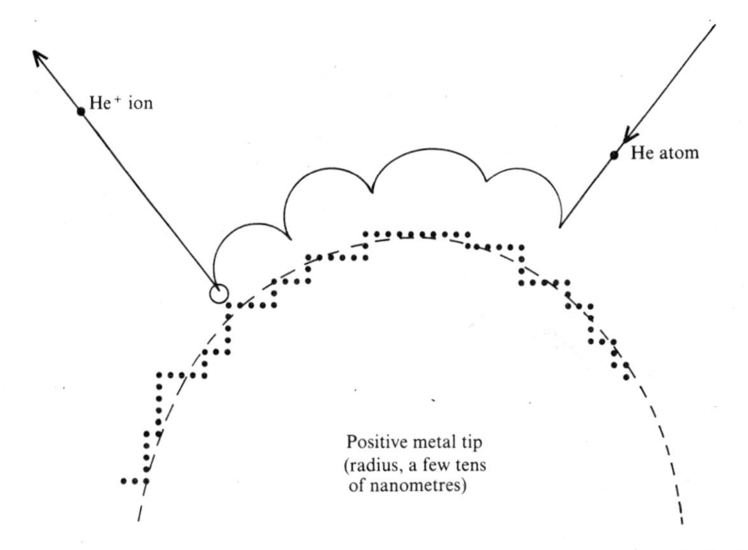

FIG. 3.18. A helium atom, polarized in the electric field, is drawn towards the tip. In a number of hops it is slowed down until it is ionized in the region above a protruding atom. The He^+ ion is then accelerated away by the field towards the fluorescent screen. Real tips do not have such simple atomic arrangements.

atoms and those circling the flat planes drawn in Fig. 3.18 can be clearly seen. Individual vacancies, emerging dislocations, and regions of disorder can be detected with the technique.

Because such a special sample geometry is required for FIM it is not possible to use the technique for surface structure analysis in the sense of studying flat single-crystal planes as used in LEED. However, an enormous amount of information about surface atomic arrangements can be derived for systems for which a tip can be formed. Adsorbates can be studied on the tip if they are sufficiently strongly bound not to desorb in the field. Some particularly beautiful observations of the clustering and migration of small groups of iridium atoms on W(110) surfaces are described in Chapter 6.

As mentioned in Chapter 2, the FIM can be combined with a mass spectrometer so as to make a device which can detect and identify an individual atom. The specimen tip is arranged to be movable so that the image of a particular atom falls on a small aperture in the flourescent screen. A short pulse of voltage is applied to the tip to cause field evaporation of the atom which passes through the aperture, and its time of flight to a single-particle detector is measured. To obtain statistically significant information on the local composition many atoms must be

FIG. 3.19. A field ion micrograph from a platinum tip of about 150 nm radius imaged with helium at 28 kV. (After Müller 1970.)

sampled but the technique is one of outstanding sensitivity. It is referred to as the atom-probe FIM and is described in detail by Müller, Panitz, and McLane (1968).

The field electron microscope (FEM) is related to the FIM. It reveals variations in work-function from place to place upon a tip and is described in Chapter 4.

Summary: Surface structure and composition

The techniques described in Chapters 2 and 3 can be used to characterize the structural and chemical state of a surface. Electron spectroscopy provides a powerful technique for establishing what species are present at the surface, and Auger electron spectroscopy is particularly sensitive if the initial-state ionizations are created by an electron beam. Sensitivities of the order of 1 per cent of an atomic monolayer are possible. Quantitative analysis is more difficult and is confined at present to systems where the surface species are in the form of uniform layers upon the substrate. Some information about the type of chemical bond present can be derived from studies of the positions and profiles of the features in the characteristic electron spectra.

The size and symmetry of the surface unit mesh of a crystallographically ordered surface can be determined with an accuracy of about 1 in 10^3 by either LEED or RHEED. Specification of the positions of each type of atom within the unit mesh is more difficult and studies of only a very small number of systems are presently available.

4. Surface properties: electronic

The earlier chapters have concentrated upon the techniques required to obtain information about what kind of atoms are present at a surface and where they are situated. As the need for other techniques arises they will be discussed, but now attention will be directed more towards the ways in which the properties of surfaces can be related to their composition and structure. Since the development of surface physics is relatively recent, most attention has been devoted to the interpretation of results obtained using electron spectroscopy, LEED, RHEED, and FIM in the study of relatively simple systems such as the low-index faces of crystals of the metal elements.

Contact potential and work function

It is convenient to define the work function of a metal in terms of the free-electron model of its electronic behaviour (e.g. Rosenberg 1974). It is the difference in energy between an electron at rest in the vacuum just outside a metal and an electron at the Fermi energy. In other types of materials, such as semiconductors and insulators, it can be regarded as the difference in energy between an electron at rest in the vacuum just outside the solid and the most loosely bound electrons inside the solid. It is clearly an important parameter in situations in which electrons are removed from a solid. For example, it affects thermionic emission, the contact potential between solids, emission of electrons in high electric fields, and the bonding of impurity atoms to a surface.

There are several physical factors contributing to the work function ϕ (see e.g. Tompkins 1967).

In the first place it can be seen, using Fig. 4.1, that the value of ϕ depends upon the depth W of the attractive potential for the conduction electrons inside the solid. This is a bulk property determined by the attraction for its electrons of the lattice of positive ions as a whole. This contribution to W will depend upon the type and arrangement of positive ions in the lattice. It is an energy of the order of a few electronvolts.

In addition, there are specifically surface contributions to W. This implies that it is wrong to draw W as independent of position in the solid, as in Fig. 4.1, but that W should change in the region of the surface. One such contribution is the *image potential*. Electrostatic theory shows that a charge $-e$ outside a conductor is attracted by an image charge $+e$

placed at the position of the optical image of $-e$ in the conducting plane. If $-e$ is a distance z from the plane the image force is thus $e^2/4\pi\epsilon_0(2z)^2$. This force is experienced by an electron escaping into the vacuum and so is a contribution to the work function. It can be shown to be negligible beyond $10^{-6}-10^{-5}$ cm away from the surface. This classical description of the image force breaks down when the escaping electron is very close to the surface ($< 0{\cdot}1$ nm away) and a quantum-mechanical description is necessary for the interaction of this electron with those remaining in the surface (e.g. Gadzuk 1972).

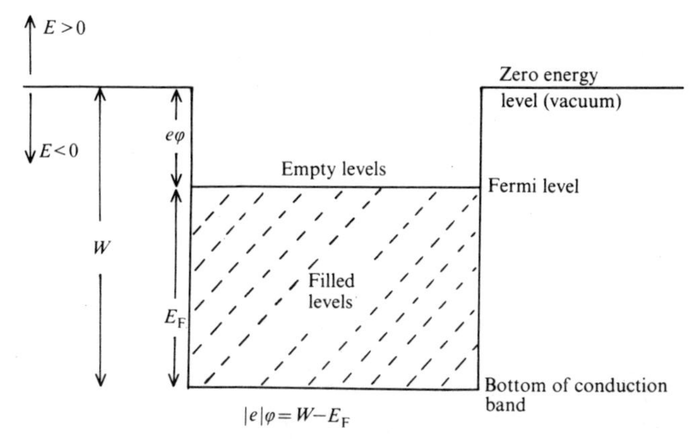

FIG. 4.1. The distribution of conduction electron energies in the free-electron model of a metal. E_F is the Fermi energy; ϕ is the work function; W is the potential well bonding the conduction band electrons into the solid.

A second surface contribution to W is the strength of a possible surface *double layer*. As the surface atoms are unbalanced, because they have matter on one side and not on the other, the electron distribution around them will be unsymmetrical with respect to the positive ion cores. This leads to a double layer of charge as indicated in Fig. 4.2. Two important effects of this double layer are that it results in the work function being sensitive to both surface contamination and the crystallographic face exposed. The contamination will affect ϕ because it will modify the double layer in a way depending upon the affinity of the contaminant for electrons. As the electron affinity of atoms depends upon their type so ϕ will vary according to the type of contaminant. Also, the orientation of the exposed crystal face will affect ϕ because the strength of the electric double layer depends upon the density of positive ion cores which in turn varies from one face to another. The contribution of the double layer to ϕ is of the order of a few tenths of an electronvolt.

These arguments lead to the expectation that the work function ϕ will depend upon which crystal face is being studied and the extent to which it is atomically clean.

The *contact potential* between two metals (e.g. Rosenberg 1974) is given simply by the difference between the work functions of the two metals.

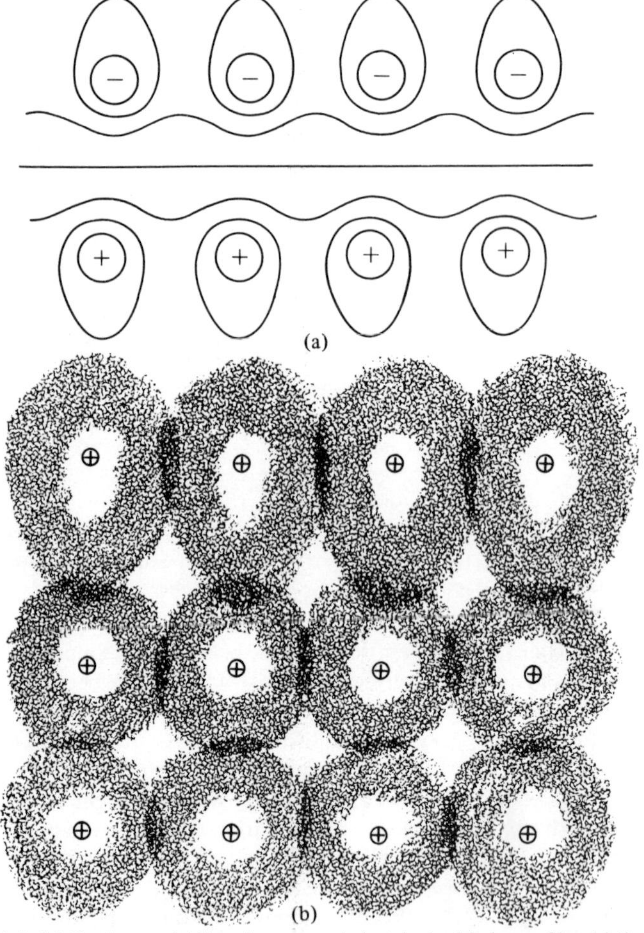

FIG. 4.2. (a) Equipotential lines in a general electric double layer. Should the centre of this layer be a surface plane there would be a potential step on passing across it. (b) A possible distribution of charge in a simple cubic lattice of positive ions. The distortion of the electron cloud at the surface causes an increased electron density between the cores compared to the bulk. A double layer with either nett positive or nett negative charge at the outside may be created by such a distortion.

The Measurement of work functions

The absolute measurement of work function is difficult. The thermionic emission of electrons depends exponentially upon ϕ through the Richardson–Dushman equation (e.g. Rosenberg 1974) and an experiment to measure thermionically emitted current as a function of temperature can be used to derive a value for ϕ. Unfortunately, measurable thermionic emission occurs only at high temperatures (ϕ is usually greater than 2 eV) and use of this technique is thus confined to those materials that have low vapour pressures at high temperatures (e.g. Ni, Mo, Pt, W, Ta). Another absolute measurement can be obtained from observations of the photoelectric yield (the total emitted photoelectron current) from the material as the frequency ν of the incident radiation is varied. It can be seen from Fig. 4.1. that, at the absolute zero of temperature, no photoemission should occur for $h\nu < |e|\phi$ and that, above the threshold ν_0 given by

$$h\nu_0 = |e|\phi\ , \tag{4.1}$$

the photoelectric yield should rise sharply as electrons at the top of the conduction band are excited into the vacuum. Fowler (1933) was the first to show how such a technique could be used to obtain accurate values for ϕ. Most clean metals have work functions of the order of a few eV and so the incident radiation for determination of the photoelectric threshold is in the ultaviolet region of the spectrum.

If a large electric field is applied to a cathode at room temperature if is possible to draw a current of electrons from it. The electrons from the metal escape to the vacuum through a thin potential barrier by quantum-mechanical tunnelling (Fig. 4.3). The size of this field emission current depends upon the work function of the cathode. The field emission microscope is based upon this phenomenon, and its use is described later in this chapter.

Two techniques which are useful for determinations of the *relative* work function are retarding potential difference (RPD) and contact potential difference (CPD) methods. A sensitive vibrating probe CPD technique is explained in Fig. 4.4. If the probe is made of a material with known work function (commonly tungsten or platinum) then the technique can be used to determine the CPD V_{12} between the probe and the sample and hence ϕ for the sample. Even if the work function of the probe is not known the technique is useful in measuring the change in ϕ of the sample as its surface is changed during an experiment. Because the technique is used to detect the null point when $V_B = -V_{12}$, it can be developed, using modern electronics, to be extremely sensitive. With care changes in ϕ of only 1 meV are detectable. In addition, it is

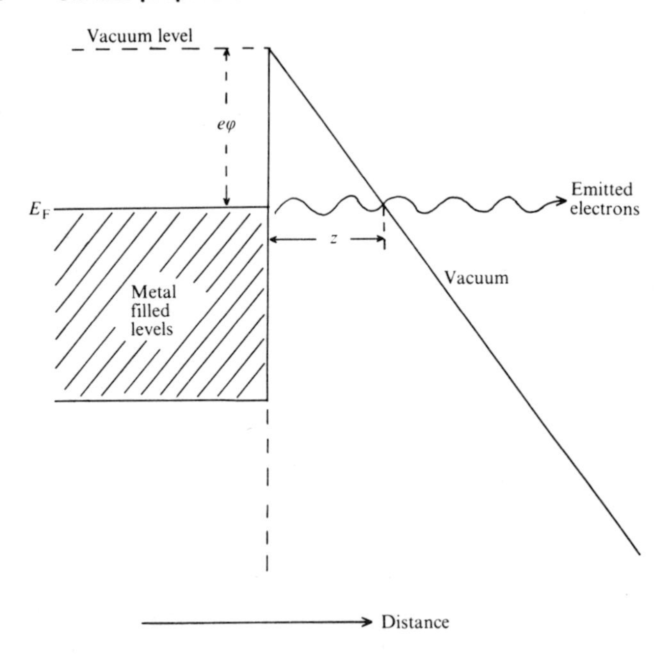

FIG. 4.3. Potential energy versus distance for a free-electron metal with its surface as one electrode in a strong electric field E. For sufficiently large E the potential barrier thickness z can be so low that electrons have high probabilities of tunnelling from near the Fermi level E_F into the vacuum.

convenient to use in conjunction with electron energy analysers and LEED optics as described in Chapter 2 and 3.

Work-function changes can be measured in the LEED apparatus of Fig. 3.13 (p. 51) using an RPD method. The sample surface is bombarded with electrons, and the total current i_c back-scattered to the screen S is measured as a function of the retarding potential $-V$ applied to the grid G2. The total potential difference acting on the electrons is $-V + V_{12}$ where V_{12} is the contact potential between the sample and the screen. If the sample surface is now changed in some way and has a new value of work function then the curve of i_c versus V will be displaced by the change $\Delta\phi$. Again this method is convenient because it is easily combined with other experiments.

Some examples of experiments involving work functions

Dependence of ϕ upon crystal face. Because of its suitability as an emitter in electronic guns, tungsten has been the subject of careful study.

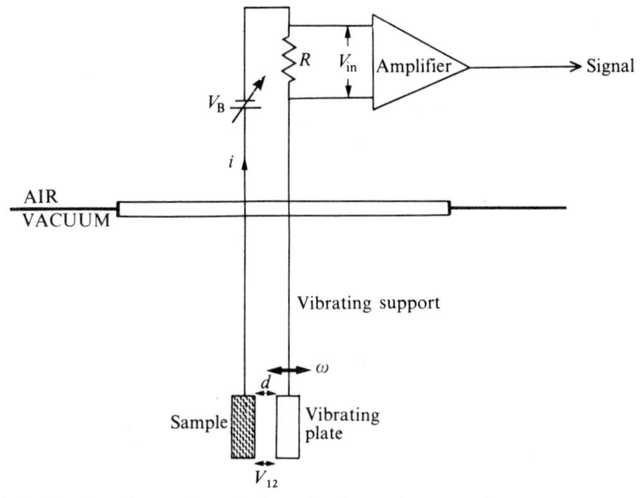

FIG. 4.4. The vibrating-probe technique for determination of relative work functions using the contact potential difference (CPD).
V_{12}, contact potential; C, capacitance; A, plate area; q, charge; d, plate separation; d_1, amplitude of vibration; ω, frequency of vibration.

$$d = d_0 + d_1 \sin \omega t \ ,$$

$$q = C(V_B + V_{12}) = (A/\epsilon_0 d)\,(V_B + V_{12})\ ,$$

$$V_{in} = iR = (dq/dt)R = -(AR/\epsilon_0 d^2)\,(V_B + V_{12})d_1\,\omega \cos\omega t\ .$$

$$\boxed{V_{in} = 0 \text{ if } V_B = -V_{12}}$$

For high sensitivity use low d, high R, and high ω.

The values of ϕ for different low-index faces of tungsten are shown in Table 4.1. The smallest values of ϕ are associated with least densely packed face. Presumably the rearrangement of electron density on the least densely packed face is such as to produce a nett positive charge on the vacuum side of the surface and so a nett transfer of electrons to the inside and thus a lowering of the work function.

TABLE 4.1

Work functions of tungsten

Crystal face	Work function (eV)
(111)	−4·39
(100)	−4·56
(110)	−4·68
(112)	−4·69

This variation of work function from one crystal face to another is demonstrated elegantly in the *field emission microscope (FEM)*. This instrument is identical in construction to the FIM described on p. 57. Now, however, no gas is admitted to the vacuum chamber and the sign of the potential on the sample tip is arranged so that electrons are accelerated out of it by a very high local electric field (typically 4×10^7 V cm^{-1}). The theory of this process is described by Müller (1970). The current leaving the region of the tip surface where the work function is ϕ is roughly proportional to $\exp(-A\phi^{\frac{3}{2}})$ and so is a very fast function of ϕ. The brightness observed on the fluorescent screen is thus a function of the value of ϕ at that place on the tip. Since, in practice, a sharp tip will be faceted so as to expose small flat areas of different crystal faces, the FEM image will consist of patches corresponding in position to these faces and of brightness depending upon the work function of each face. An FEM image of a tungsten tip is shown in Fig. 4.5, which is labelled with the Miller indices of each facet giving a bright spot. The brightest spots correspond to the lowest values of ϕ in Table 4.1.

The changes in ϕ produced by adsorbed atoms or molecules can be followed in the FEM and the diffusion of these adsorbates over the tip surface followed. Spatial resolutions of about 2·5 nm are possible.

Dependence of ϕ upon contamination. If the adsorption of atoms or molecules results in the transfer of charge to or from a surface then the work function will change. The adsorbed species may be polarized by the attractive interaction with the solid surface, or, more extremely, it may be ionized. If it is polarized with the negative pole towards the vacuum the consequent electric fields will cause an increase in work function. This can be seen by considering the effects of a deeper positive potential at the edge of the well in Fig. 4.1. Conversely, if the positive pole of the polarized adsorbate is outwards then the work function of the substrate will decrease. Similarly, an adsorbate which is in the form of positive ions will have transferred electrons to the substrate and so decreased its work function. So long as the number of adsorbed ions (or polarized atoms or molecules) is low enough for interactions between them to be negligible, the change in work function will be proportional to the number of ions adsorbed. The variation of work function with type of adsorbed species is discussed by Somorjai (1972).

The effects of an adsorbate upon work function are clearly demonstrated in the work of Adams and Germer (1971), who studied the adsorption of molecular nitrogen upon W(100), (310), and (210) surfaces. By correlating work function and LEED observations with measurements of the amounts of nitrogen desorbed into a mass spectrometer (flash desorption) after each experiment they were able to obtain the data

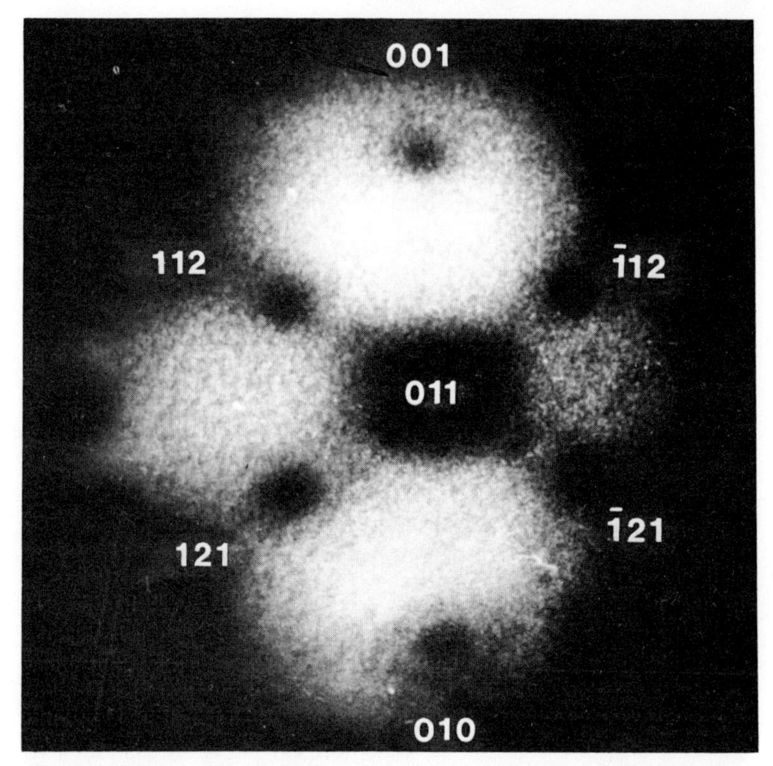

FIG. 4.5. An FEM image of a (011) single-crystal tungsten tip of radius 210 mm. The patches in the image are labelled with the Miller indices of the crystal facets to which they correspond. (By courtesy of Dr. H. Montagu-Pollock, University of Lancaster.)

shown in Fig. 4.6. By studying the variation of the LEED patterns with the dose (rate of arrival of molecules x time of exposure) they were able to conclude that adsorption ceased after half a monolayer of nitrogen had arrived. This *saturation coverage* was removed by heating the crystal quickly to desorb the nitrogen. Some of the nitrogen is collected by a mass spectrometer the output of which is then calibrated in terms of the known saturation coverage. Thus the data for the abscissa of Fig. 4.6 can be obtained. The interpretation of the changes $\Delta\phi$ in terms of the distribution of electrons at the surface is not clear at present.

Surface states and band-bending

In a bulk solid, the potential due to the array of ion cores in crystallographic sites will vary in a three-dimensionally periodic way. If solutions of Schrödinger's equation are sought for a periodic potential it is found that no travelling-wave solutions exist for particular values of the wave-vector **k** of an electron. The absence of eigenstates for these values of **k** leads to the band gaps in the electronic structure of the solid. The derivation of this result is given, for example, in Rosenberg (1974). The centre of the band gap corresponds to the condition for elastic scattering of electrons by the periodic potential.

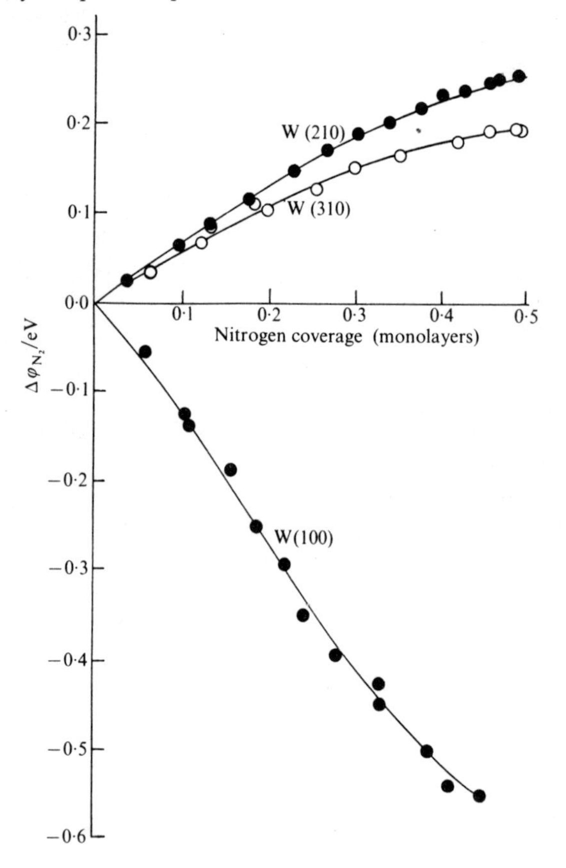

FIG. 4.6. Variation of change in work function $\Delta\phi$ of various tungsten single faces with coverage of nitrogen. (After Adams and Germer 1971.)

At the surface this three-dimensional periodicity is lost, and travelling-wave solutions of the Schrödinger equation can be found which are situated within the band gap of the bulk solid. These special solutions are waves which can travel parallel to the surface but not into the solid.

Thus they are localized at the surface and can have energies within the band gap of the bulk band structure. These solutions are known as *surface states*. If charge (either electrons or holes) is situated in these surface states it will result in electrostatic fields penetrating into the solid, and thus a varying electrostatic potential on passing from the surface to the interior. This varying potential distorts the band structure of the bulk solid – an effect which is called *band-bending*.

Surface states can be associated not only with the termination of a three-dimensional potential at a perfect clean bulk exposed plane but also with changes in the potential due to relaxation, reconstruction, structural imperfections (such as emerging dislocations), or adsorbed impurities. If the charge associated with any of these surface states is different from the bulk charge distribution then band-bending will result. The surface states will affect the electrical properties of a surface by acting as a source (or a sink) of electrons and the chemical reactivity by modifying the affinity of the surface for electrons.

Quantitative estimates of the effects upon band shapes of charge localized in surface states are quite difficult to make. If the free-carrier density in the material is very high then compensation of the charge

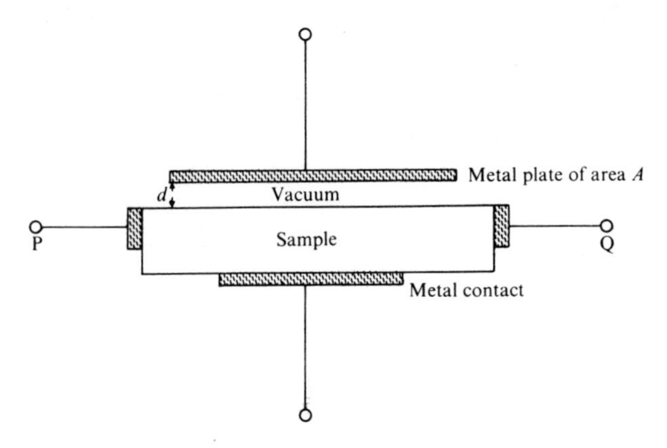

FIG. 4.7. A capacitor experiment which can be used to demonstrate the existence of surface states and the changes in charge trapped in them with varying surface conditions. Total capacitance C, surface capacitance C_s; geometrical capacitance C_g.

$$\frac{1}{C} = \frac{1}{C_s} + \frac{1}{C_g},$$

$$C_s = \frac{\epsilon_0 A}{\epsilon_0 A - Cd} \cdot C$$

at the surface by flow of carriers out of the bulk is effective (the conduction-band electrons 'screen' the surface charge). Thus, metals tend to have negligible band-bending. In semiconductors and insulators such complete compensation is not possible because of their much lower free-carrier densities, and the existence of occupied surface states can result in surface potentials of the order of millivolt or volts, the band-bending effects extending some micrometres into the solid. The simplest theories for surface states are outlined, for instance, in Wert and Thomson (1970) and for band-bending in semiconductors by McKelvey (1966).

If the potential varies on passing from the surface to the interior of a solid then the surface must have a capacitance. Measurement of the surface capacitance is one of the possible probes of the electronic properties of a surface. If a parallel-plate capacitor (Fig. 4.7) is constructed such that its geometrical capacitance is small compared to the surface capacitance C_s then changes in C_s can be measured as a function of applied electric fields, coverage of various adsorbates, or exposure to electromagnetic radiation. The electronic properties of the surface can be inferred from the results of such experiments. Alternatively, the *surface conductivity* can be measured using contacts P and Q on the sample of Fig. 4.7 when charge is induced at the surface by applying a potential to the metal plate. If there were no surface states the charge would be injected into the conduction band of the sample and the change in conductivity expected is predictable. In fact, only about one-tenth of this expected change is found, suggesting that relatively immobile charge is trapped in surface states.

Surface states can also be observed by the techniques of electron spectroscopy described in Chapter 2. Semiconductor surfaces are particularly interesting in this respect because it is anticipated that they may show reconstruction.

An important example of a reconstructed surface is found for Si(111). The LEED pattern of a clean Si(111) surface (Fig. 3.14(b), p. 53) shows extra features at positions separated by one-seventh of the distance between the spots due to a bulk exposed plane and thus atomic disturbances must be occurring in the surface with 7 times the periodicity of the bulk atomic arrangement. The pattern is thus referred to as Si(111)7. A silicon atom in the bulk is covalently bonded to four nearest neighbours in a diamond cubic structure. At the (111) bulk exposed plane each atom (Fig. 4.8) has one covalent bond 'dangling' into the vacuum in the direction of the surface normal. Such a dangling bond is energetically unfavourable, and the surface may be able to reduce its free energy by reconstructing in such a way as to reduce the number and/or energies in these bonds. This is believed to be the case for Si(111)7, but no structure analysis for such a complex diffraction pattern has yet been possible.

If a surface has dangling bonds *or* reconstructs so as to reduce the energy associated with the surface, then the bonding for surface atoms will be different from the bonding for bulk atoms and surface states will exist. If electrons are energy-analysed (by one of the techniques in Chapter 2) after being scattered from the surface, some of them will have lost specific amounts of energy in exciting particular processes in the solid. These processes are mentioned on p. 22 and show up as features labelled as losses in Fig. 2.2(c), p. 15). The detailed study of such loss spectra is often referred to as *electron-loss spectoscopy*. As the primary energy is

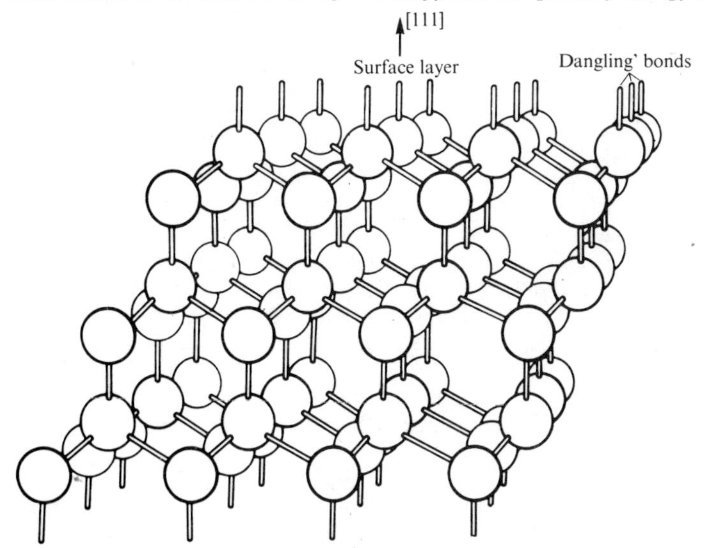

FIG. 4.8. 'Dangling' bonds from the (111) surface of a covalently bonded diamond cubic structure.

varied the losses move along the energy scale fixed by a particular energy difference from the elastic peak. The largest features in most loss spectra are the surface and bulk plasmon losses but many other features can be observed which are due to excitation of an electron in the solid form its ground state to some empty state or band above the Fermi level. A careful study of the loss spectra of various Si(111) and Si(100) surfaces. (Rowe and Ibach 1973) has shown that three features in the electron spectrum corresponding to losses of approximately 2 eV, 8 eV, and 14 eV can be associated with surface states. The correlation with surface states is made by noting the following: the way that oxygen adsorption decreases the size of the 14 eV feature, presumably because the oxygen atoms are using up the dangling bonds; the way the 8 eV peak is present only for well-ordered surfaces (as judged by LEED); and the way the strength of the loss features depends upon the surface roughness induced by Ar^+ ion bombardment. Thus, surface states, identified by peaks in loss spectra, can be correlated with particular bonds between silicon surface atoms. It is too soon to say if this method can be applied to other solid surfaces.

Plasmons

The loss processes described on p. 22 sometimes have different values for surface and bulk environments. The losses described above are single-particle excitations of electrons out of the bonds of surface atoms into empty states just above the Fermi level. They are also surface excitations because they are specific to the special bonds which exist at the surface. Two other important kinds of excitation are into *plasmons* and *phonons* which are respectively collective excitations of the electrons and the atoms in the solid. Phonons will be discussed in Chapter 5.

A plasmon is quantized oscillation in the density of an electron gas. Such oscillations can be excited by shooting in a charged particle or a photon. The Coulomb field of the former and the electromagnetic field of the latter cause a redistribution of charge in the electron gas which launches plasma oscillations. The simple theory of plasmons in a bulk solid is described in many books on solid-state physics (e.g. Rosenberg 1974; Kittel 1967). The energy of a plasmon is related to the density n of a free-electron gas by

$$E = \hbar\omega_p = \hbar\,(ne^2/\epsilon_0 m)^{\frac{1}{2}}. \tag{4.2}$$

A metal with conduction band electron densities in the range $10^{27} - 10^{29}$ electrons m^{-3} will thus show plasmon losses of the order of a few electron volts. The exciting particle may generate more than one plasmon as it passes through the solid and so suffer multiple plasmon losses. Magnesium metal exhibits such multiple plasmon losses as is shown in Fig. 4.9.

If a semi-infinite electron gas is terminated in a plane surface Laplace's equation can be solved for possible charge density fluctuations and plasma oscillations predicted which are periodic in the plane of the surface but decay away exponentially into the electron gas (e.g. Kittel 1967). The quantized unit of these oscillations is called a *surface plasmon*. By applying the electromagnetic boundary conditions (the tangential component of the electric field and the normal component of the displacement must both be constant across the surface–vacuum interface) and using the dispersion relationship for the dielectric constant of an electron gas (again, see Kittel 1967) it can be shown that the frequency ω_s of a surface plasmon is related to ω_p by

$$\omega_s^2 = \frac{1}{2}\,\omega_p^2\,. \tag{4.3}$$

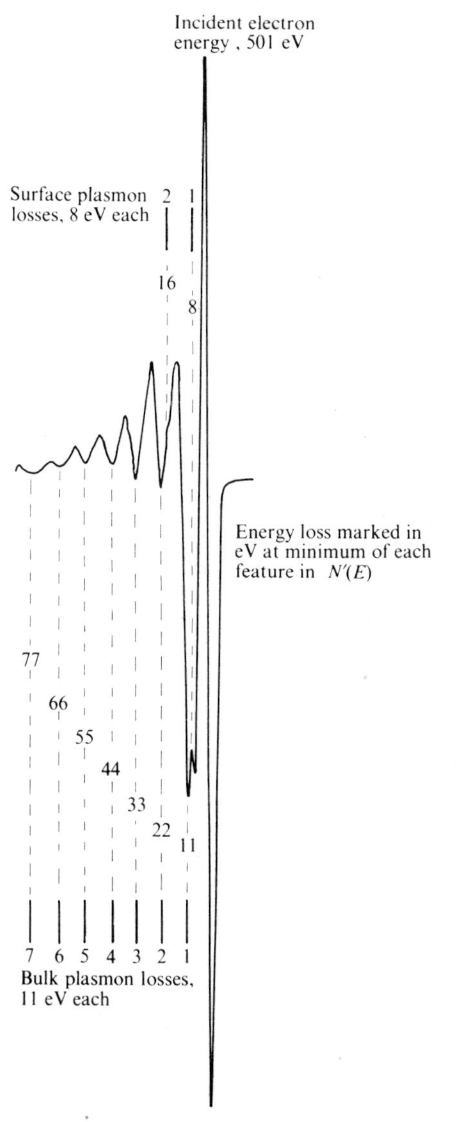

FIG. 4.9. The electron-loss spectrum from a clean surface of polycrystalline magnesium using a beam of incident electrons of energy 501 eV. Multiple bulk plasmon losses of 11 eV and multiple surface plasmon losses of 8 eV are clearly visible. (By courtesy of Dr A. P. Janssen.)

The surface plasmon excitations in magnesium can be seen at intervals of 7·1 eV in Fig. 4.9.

The relative intensities of bulk and surface plasmon excitations will depend, for instance, upon the energy and angle of incidence of a primary electron beam and upon the energy, angle of incidence, and state of polarization of an incident electromagnetic wave. For incident electrons the surface plasmon becomes more pronounced as the primary energy is reduced and as the angle of incidence is increased, because in both cases the penetration of the electrons decreases and so does the probability of exciting bulk plasmons. Of course, if the primary energy falls below the surface plasmon energy no collective excitations of the electron gas are possible.

A detailed study of plasmon-loss processes may help to obtain some understanding of the distribution of charge at a surface. The work function is a gross property in that it is a single parameter whose value is determined by a number of independent microscopic properties of the solid and its surface. Just as in bulk solid-state physics, the densities of available and of occupied states for electrons and the relationship between electron energy and momentum (*the dispersion relation*) are useful concepts in describing surface electronic properties.

One means of obtaining an idea of surface charge distributions is to attempt to find the dispersion relationship for surface plasmons. A promising method for doing just this is called *inelastic low-energy electron diffraction* (ILEED). The apparatus used to perform an ILEED experiment can be the same as the LEED apparatus of Fig. 3.13 (p. 51) but the potentials and electronics are arranged as in Auger spectroscopy, so that electrons that have suffered a selected energy loss ΔE reach the screen and are detected. Experimentally the variation of measured collected electron current with primary beam energy, angle of incidence ϕ, azimuthal angle ψ, angle of emitted electron beam ϕ', and energy loss ΔE is observed. This rather complex data can then be compared with models for the scattering process in the solid. If processes which involve a plasmon loss before elastic diffraction and a plasmon loss after elastic diffraction are both taken into account it is possible to find the dispersion relations for both bulk and surface plasmons. This approach is still in the early stages of its development but has been applied with some success to Al(111) surfaces by Porteus and Faith (1973).

Surface optics

When plane-polarized light is reflected from a solid surface the reflected wave is, in general, elliptically polarized. The ellipticity can be measured using standard techniques of polarimetry (Beshara, Buckman, and Hall 1969). The experimental results can be expressed in terms of the amplitudes of two components of the oscillating electric field in the reflected light and the phase difference between them. The measurement and interpretation of this ellipticity as a function of the angle of incidence, the plane of polarization of incident light, and its wavelength forms the subject of *ellipsometry.* It is a subject with a long history, since Malus detected polarization by reflection from metals in 1808, but it has been plagued by two substantial difficulties which could be removed only in the recent past.

The first difficulty arises because the solution of Maxwell's equations for reflection at a real (absorbing) surface leads to complicated expressions involving the refractive index n_c, the wavelength, and geometrical parameters. The refractive index itself is a complex quantity involving a real refractive index n and an imaginary term due to absorption and usually called the extinction index k.

$$n_c = n - \mathrm{i}k\ . \tag{4.4}$$

The values of n and k, and their dependence on wavelength, are interesting physical parameters in that they are related to the response of the electrons in the solid to the electric field of the incident light wave. The real part of the refractive index is determined largely by the extent to which atoms in the solid are polarized by this electric field. The imaginary part is determined by the energy lost from the incident wave to single particle and collective excitations of the electrons. Thus, the optical properties are intimately related to the plasmon dispersion relations mentioned above. However, the parameters actually measured in an ellipsometric experiment are related only through complicated equations to calculable quantities like the reflectivity and phase change Δ on reflection. In turn, these quantities are related through a second set of complicated equations to the values of n and k at any particular wavelength. The solution of this difficulty (which varies from simply tedious to quite hideous – see e.g. the review by Heavens (1964)) awaited the development of the modern electronic computer.

The second difficulty arose from the very high surface sensitivity of the ellipsometric method. UHV techniques and the methods of surface analysis described in Chapters 2 and 3 were required before reproducible results could be obtained and interpreted. Vrakking and Meyer (1971) have shown how the size of an Auger electron peak and the change $\delta\Delta$ in the phase change on reflection due to adsorbed layers of O_2, H_2S, and CH_3SH on Si(100) are linearly related up to at least monolayer coverage (depending upon wavelength). Observable changes in Δ are possible in these systems for as little as 5 per cent of a monolayer of coverage by the adsorbed species.

Electron spin resonance

Electron spin resonance spectroscopy continues to be a powerful technique for probing the bonding and environment of atoms in bulk solids. The spectrometer is designed to measure the adsorption of energy from an electromagnetic field due to changes in the spin states of unpaired electrons. Resonance occurs at a frequency ν for a sample in a magnetic field H when

$$h\nu = g\beta H \, , \tag{4.5}$$

where g is the spectroscopic splitting factor and β is the Bohr magneton. If the orbital energy levels are split by a crystal field then the size and symmetry of g will depend upon the magnitude and symmetry of the crystal field.

The technique has been applied to the study of surface atomic species (e.g. Lunsford 1972) but, in order to obtain sufficient sensitivity, it has been normal practice to study adsorption upon the surfaces of a collection of very small particles. Although increasing the surface area this approach is almost inevitably confined to conventional vacuum techniques (as opposed to UHV) due to trapped gases. In addition, it gives results which are some kind of sum over the effects of many different surface orientations. Nevertheless, the method is so powerful in bulk chemical investigations that means may be found to extend its application to clean surface studies.

Summary

The wavefunctions of the electrons of a solid are expected to be different at the surface as compared to the interior. This difference is important in a variety of contexts – e.g. it affects the manner in which an adatom bonds to the surface (Chapter 6), the emission of electrons into the vacuum, and the details of electron diffraction processes (Chapter 3). Techniques for the detailed evaluation of surface densities of states and dispersion relations for surface excitations are still in their infancy. The work function and the ellipticity of reflected light are both very sensitive to small changes in the electronic states at the surface but are difficult to interpret in terms of atomic models. ILEED and electron-loss spectra show features characteristic of surface states, and it may be possible to develop these methods so as to give detailed information on surface electronic behaviour.

5. Surface properties: atomic motion

Up to this point the discussion of both methods and properties has been in terms of rigid lattices of atoms or molecules. In practice, of course, the atoms are in motion, and this motion should be included in a treatment of any properties it may affect. In this chapter the atomic vibrations at a surface will be discussed, as they reveal themselves in the temperature dependence of the intensity of a LEED pattern and in special features in electron-loss spectra. More extreme atomic motions occurring during the melting of a surface and during the motion of an adatom across it will also be discussed.

Surface lattice dynamics

It is well known that there is a fall in the intensity of the beams in an X-ray diffraction experiment (using a bulk single crystal) as the crystal temperature is raised. At the same time the intensity in the diffuse background of the diffraction pattern becomes higher. The simplest explanation of these observations is that the individual atoms of the crystal are vibrating independently about their equilibrium positions and, as a result, the exact Bragg condition is not met. This is because scattered waves from the rigid lattice that were adding up in phase now have phase differences fluctuating with time due to the motion of the scatterers. The effect of this motion upon the intensity of the elastically diffracted beams is described in many textbooks (e.g. Kittel 1967). If I_0 is the intensity elastically scattered into a beam by a rigid lattice then the intensity I_g due to elastic scattering by a vibrating lattice in the direction determined by Bragg scattering due to a reciprocal-lattice vector $\mathbf{g}$ is given by

$$I_{\mathbf{g}} = I_0 \exp(-\alpha \langle u^2 \rangle |\mathbf{g}|). \tag{5.1}$$

In deriving this equation it is assumed that the atoms are in simple harmonic motion. $\langle u^2 \rangle$ is the mean-square amplitude of vibration in the direction of $\mathbf{g}$ and α is a constant whose value depends upon the number of dimensions in which the atoms are allowed to vibrate. If oscillation in one dimension along $\mathbf{g}$ is chosen as a model than $\alpha = 1$; if oscillation in three dimensions is chosen, $\alpha = \frac{1}{3}$. The exponential factor of eqn (5.1) is usually called the Debye-Waller factor and is often written as $\exp(-2M)$.

The same kind of effect is observed in LEED, but because LEED intensities arise from the first few atomic layers of a crystal the appropriate value of $\langle u^2 \rangle$ is that for the surface atoms. Indeed, as the energy of the incident electron beam is raised, the penetration increases and the relevant number for $\langle u^2 \rangle$ changes from a surface to a mainly bulk value. Because of the absence of nearest neighbours on the vacuum side it is likely that $\langle u^2 \rangle$ at the surface will be greater than that in the bulk. Further, this very asymmetry in the potential around the surface atoms is likely to require an anharmonic description of the lattice vibrations. This must be particularly true of the component of vibration normal to the surface $u_{\mathbf{n}}$.

By using the Debye model of the solid as a three-dimensional elastic continuum it is possible to derive theoretical expressions for $\langle u^2 \rangle$ in eqn (5.1) (Kittel 1967). This can be inserted in eqn (5.1) and a new equation derived which relates the observed intensity to other, measurable quantities. A beam of wavelength λ incident upon the surface of an angle ϕ is scattered into an (00) beam whose temperature-dependent intensity $I_{00}(T)$ is given by

$$I_{00}(T) = I_{00}(0) \exp \left\{ \left| -\frac{12h^2}{mk} \left(\frac{\cos \phi}{\lambda} \right)^2 \frac{T}{\Theta^2} \right| \right\}. \qquad (5.2)$$

In this equation $I_{00}(0)$ is the specularly reflected intensity from a rigid lattice, h is Planck's constant, m is the atomic mass, k is Boltzmann's constant, T is the temperature, and Θ is the Debye temperature. †
If the intensity of a specular LEED spot is measured at constant λ and ϕ as a function I then a plot of $\log \{I_{00}(T)\}$ versus T should be a straight line and eqn (5.2) can be used to derive a value for Θ. MacRae's (1964) data for Ni(110) is plotted in Fig. 5.1, which shows only his results for the specularly reflected beam. At the energy of 35 eV used to produce these results the value of Θ found from the slope of the line is 220 K. The bulk Debye temperature of nickel is 390 K, and so this result suggests that the atoms in the layers penetrated by the 35 eV electrons (the top one or two layers) have higher values of $\langle u^2 \rangle$ than do the bulk atoms. Some results of this kind for a few systems are listed in Table 5.1. The values of $\langle u^2 \rangle$ derived from such experiments depend upon the accuracy

†The Debye temperature is a characteristic of the bulk solid which is oftern first encountered in a discussion of the heat capacity. It is associated with energy of the highest frequency ($\omega_{\mathbf{max}}$) phonon mode possible in the Debye model of vibrations in the solid,

$$h\omega_{\mathrm{max}} = k\,\Theta$$

(Kittel 1967).

TABLE 5.1
Some surface vibrational data derived from LEED observations

Material and surface	Reference	Θ (surface) (K)	Θ (bulk) (K)	$\langle u_n^2\rangle$ surface / $\langle u_n^2\rangle$ bulk
Xe(111)	Tong (1973)	30–5	43	3·5–2
Bi(0001)	Goodman (1970)	48	116	2·4
Ni(110)	MacRae (1964)	220	390	1·8
Cr(110)	Kaplan (1971)	333	600	1·3

$\langle u_n^2\rangle$ is the mean-square vibrational amplitude normal to the surface.

with which multiple-scattering effects are included in calculating the LEED intensities and the force law used to describe the interaction between the atoms. Because Tong, Rhodin, and Ignatiev (1973) chose to study Xe(111) surfaces, which show nearly kinematical LEED intensities, they could examine the effects of choosing different force laws, and it is this range of choices which gives the spread of values to Θ and $\langle u_n^2\rangle$ in Table 5.1.

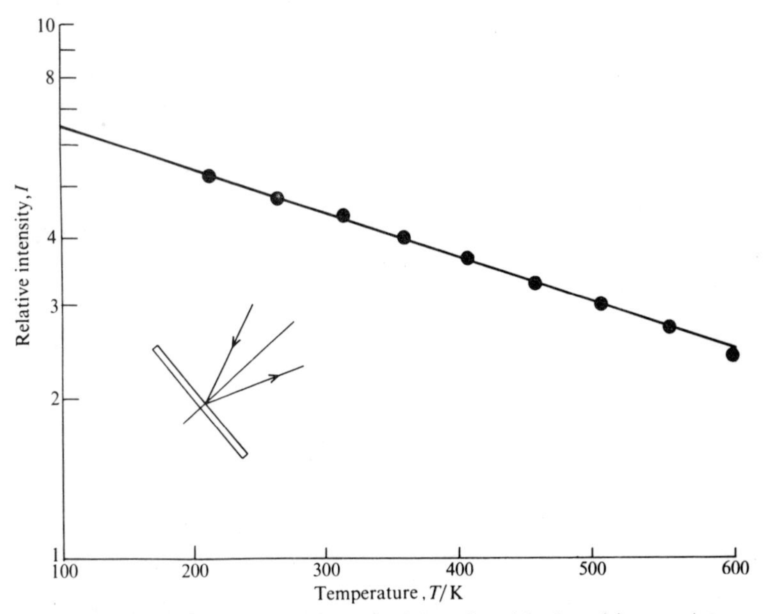

FIG. 5.1. A plot of the measured intensity I (on a logarithmic scale) versus temperature T for the specular LEED spot of Ni(110) at 35 eV. The slope yields a value of 220 K for the Debye temperature. (After MacRae 1964.)

Another result of the change in the number of nearest neighbours on moving from the bulk to the surface is that different modes of vibration of the lattice become possible. The vibrational states of the bulk crystal are quantized, each quantum being called a phonon. The vibrational properties of the bulk solid are described by using a phonon dispersion relation $E(q)$ which expresses the energy E of a phonon as a function of its wavevector magnitude q. Just as there can be gaps within which particular values of k are not allowed in the electron dispersion relation $E(k)$ of a solid, so there can be gaps in $E(q)$ for the phonons. For electrons the gaps are due to $\mathbf{k}$ satisfying the condition for elastic scattering by the periodic potential of the solid. For phonons the gaps occur when q lies on a Brillouin zone boundary because atomic vibrations with wavelengths shorter than the interatomic spacing are not possible (see e.g. Rosenberg 1974).

At the surface there are vibrational modes which are not allowed within the solid. They are analogous to the surface plasmons in that they propagate with wavevector $\mathbf{q}$ parallel to the surface and decay exponentially away in amplitude along the surface normal. These modes are called surface phonons.

The two most widely used techniques for observing the vibrational spectra of solids are infared spectroscopy and inelastic neutron scattering. The latter is particularly powerful in that both the energy and the momentum of a neutron which has lost energy in exciting a phonon can be measured. By varying the sample orientation and the neutron detector position the function $E(q)$ can be determined. However, the neutron – phonon scattering cross-section is very weak, and so the technique has not been applied to surface problems. In infrared spectroscopy the electric field of the incident infrared beam can couple to the phonon modes of the solid and the resultant absorption of the beam can be measured as a function of the energy of incident beam. Spectrometers of this type are made to function over the energy range 1·5 meV to 0·1 eV (wavenumbers of 12 – 800 cm^{-1}). Sufficient sensitivity for surface modes to be detected can be obtained with infrared techniques if the sample is in the form of a large number of small particles. The high surface–volume ratio has thus allowed the detection of surface phonons in materials like magnesium oxide. However, such small-particle experiments are difficult to interpret, as many crystal faces are present on the particles and contamination by impurities is a strong possibility (Chapter 1). Some applications of infrared spectroscopy to surfaces are described by Amberg (1967).

The excitation of surface phonons by low-energy electrons, used in UHV conditions and with single-crystal surfaces has been demonstrated by Ibach (1972). He has used very high energy resolution electron-loss spectroscopy (Chapter 4). The high resolution is essential because phonon losses are small (a few tens of millielectronvolts) and are normally submerged in the broad energy distribution of electrons emitted from a conventional electron gun. This difficulty is overcome by using a dispersive analyser and an electron gun together to produce a monochromatic beam. After scattering from the solid surface the electrons are energy-analysed with a second dispersive analyser. The energy-loss spectrum from a UHV cleaved Si(111) (2 x 1) surface structure obtained by Ibach is shown in Fig. 5.2.

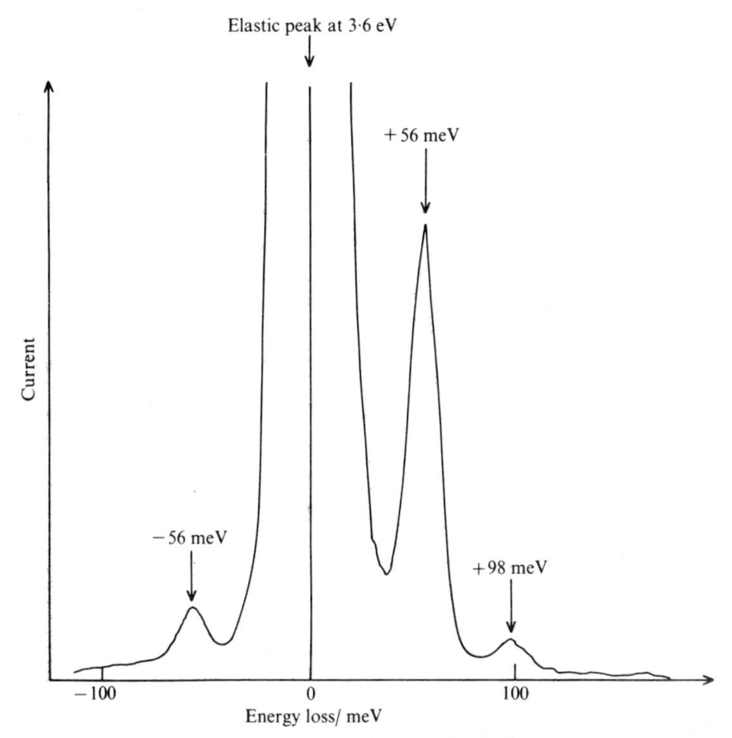

FIG. 5.2. Energy-loss spectrum from Si(111)(2 x 1) using a monochromatic primary beam of energy 3·6 eV. The plane of incidence contains one of the {011} directions. The small peak at 98 meV is attributed to a contaminant. A loss at +56 meV and a smaller gain at −56 meV are visible. (After Ibach (1972, p. 717, Fig. 8).)

Although sophisticated techniques are required to observe these energetically small surface phonon losses they offer a potentially important means of obtaining information about the surface. As the surface phonon energy depends upon the type, the bonding, and the crystallography of the surface atoms; comparison between such observations and the energies calculated from models of the surface should help to determine these parameters. For instance, if a LEED structure determination (Chapter 3) predicts a particular geometry of surface atoms it should be possible to calculate the surface phonon energy for simple directions within that structure and compare this with the observed energy loss. This would provide an independent test of the proposed structure.

Surface diffusion

As described above, at any finite temperature the atoms at the surface of a perfect crystal are vibrating at some frequency ν_0. Thus ν_0 times a second each atom strikes the potential-energy barrier separating it from its nearest neighbours. Sometimes (Fig. 5.3) the thermal energy

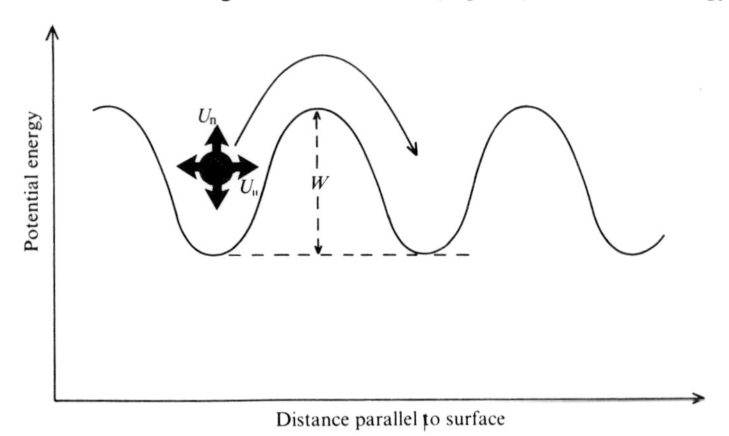

FIG. 5.3. An atom A vibrating with amplitudes u_n and $u_{\parallel}$ normal and parallel to simple sinusoidal potential energy surface. The probability that it will surmount the barrier of height W and reach another potential-energy minimum is porportional to exp $(-W/kT)$.

fluctuations give the atom sufficient energy for it to leave its initial site in the surface and become an adatom in a neighbouring position of potential-energy minimum. This is the simplest picture for the start of self-diffusion of an atom across a perfect surface. In practice, a real surface will contain many defects on this atomic scale (Fig. 5.4), and so there will be many different sites for surface atoms. The frequency ν with which an atom will escape from a site will depend upon the height W of the potential energy barrier it has to surmount during the escape,

$$\nu = z\nu_0 \exp(-W/kT). \tag{5.3}$$

In this equation z is the number of equivalent neighbouring sites for the atom. This equation can be combined with a mathematical treatment of an atom executing a random walk for a time t over a mean-square distance $\langle R^2 \rangle$ to give

$$\langle R^2 \rangle = Dt \tag{5.4}$$

and

$$D = D_0 \exp(-W/kT) . \tag{5.5}$$

D is normally referred to as the diffusion coefficient and is usually expressed in square centimetres per second. A detailed description of the thermodynamics and the diffusion theory leading to these results can be found, starting with Blakely (1973).

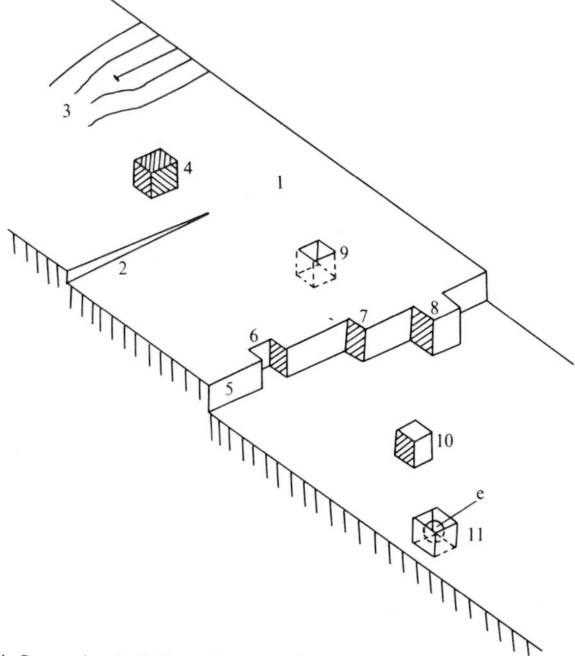

FIG. 5.4. Some simple defects that are often found on a low-index crystal face. 1. The perfect flat face itself – a terrace; 2. an emerging screw dislocation; 3. the intersection of an edge dislocation with the terrace; 4. an impurity adatom (adatoms are discussed in Chapter 6); 5. a monatomic step in the surface – a ledge; 6. a vacancy in the ledge; 7. a step in the ledge – a kink; 8. an adatom of the same kind as the bulk atoms situated upon the ledge; 9. a vacancy in the terrace; 10. an adatom on the terrace; 11. a vacancy in the terrace where an electron is trapped – in an alkali halide this would be an F-centre.

Because of the variety of different sites involved on a real surface a surface diffusion experiment usually gives a measure of W which is an average over several different diffusion processes. Some processes which might be expected to occur on the surface shown in Fig. 5.4 are as follows :

1. a single adatom (10) may hop across a terrace in jumps several lattice constants long;
2. an adatom (8) may diffuse along the length of a ledge (5);
3. a vacancy (9) may diffuse about by being successively filled by surface atoms.

Of course, more complex processes may also occur and it is part of the objective of a diffusion experiment to draw conclusions about mechanism of diffusion by comparing observed values of D with those calculated for various mechanisms.

Some particularly elegant observations of the diffusion of single adatoms upon terraces upon different tips in a field ion microscope (Chapter 3, p. 57) have been carried out by Bassett (1973). He could observe $\langle R^2 \rangle$ by vapour-depositing a single adatom on to a tip and comparing micrographs before and after an experiment such as heating the tip to a known temperature for a known time. Values of D for each temperature T could then be derived from eqn (5.4) and plots of $\log D$ versus $1/T$ used to derive values of W, the activation energy for surface diffusion. Some of the results are summarized in Table 5.2. It can be seen that

TABLE 5.2

Activation energies for surface diffusion of single adatoms as observed by FIM

Adatom \ Terrace	W (011)	W (112)	W (321)	Ir (111)	Ir (113)	Rh (111)	Rh (113)
W	0·87	0·57	0·84	–	0·99	–	–
Re	1·04	0·88	0·88	0·52	1·17	–	–
Ir	0·78	0·58	–	–	0·92	–	–
Pt	~0·6	–	–	$<0·41$	0·69	–	–
Rh	–	–	–	–	–	0·24	–

Activation energies W are in eV.
Data from: Bassett (1973) and Bassett and Parsley (1970).

there are considerable variations in W from one crystal face to another even for the same type of atom on the same substrate material. Also, the diffusion can be very anisotropic on some planes – e.g. the adatom motion is along the natural channels in the W(112) and Ir(113) surfaces. Although not shown in Table 5.2, the pre-exponential factor D_0 also varies from face to face and between materials – between $3{\cdot}8 \times 10^{-7}$ $cm^2 s^{-1}$ for tungsten adatoms on W(211) and $1{\cdot}5 \times 10^{-2}$ $cm^2 s^{-1}$ for rhenium adatoms on W(110).

Theoretical treatments of the diffusion of single adatoms on clean metal surfaces are difficult in that it appears to be necessary to include both the relaxation between substrate atoms and the adatom and also the variation that can occur in the bond strength per bond when the number of nearest neighbour atoms is varied if agreement between theory and experiment is to be obtained.

Other experimental methods of studying surface diffusion do not involve the observation of single atoms but the change in shape of gross features on the surface by mass transport. One powerful technique of this kind uses a measurement as function of time and temperature of the amplitude of a sine-wave topography etched chemically into a crystal surface. The amplitude of the surface roughness can be measured *in situ* in UHV by allowing light from a laser to be diffracted by the sine wave and measuring the intensity distribution in the diffracted beams. The technique is explained by Blakely (1973). The diffusion coefficients and binding energies measured with this method are different from those obtained from the FIM observations. The latter involve the diffusion of single adatoms over a surface chosen to be free of ledges, kinks, vacancies, and impurities. The former may have at least the first three kinds of defects and, as a result, the measured activation energies of the individual processes and the populations of each type of defect taking part. Nevertheless, the mass-transport techniques are of interest in so far as they help to give understanding of technologically important processes such as sintering and creep.

Surface melting

As a piece of bulk crystal is heated up its atomic vibrations become stronger and stronger until a temperature is reached at which the crystallographic order is lost and the atoms are in a disordered conglomeration. This order–disorder transition is melting. A two-dimensional array of atoms might be expected to melt at a lower temperature than a bulk lattice because of the smaller number of nearest neighbours in the former. Although it is not possible to study a two-dimensional layer of atoms in isolation it is possible to grow two-dimensional layers of atoms of one

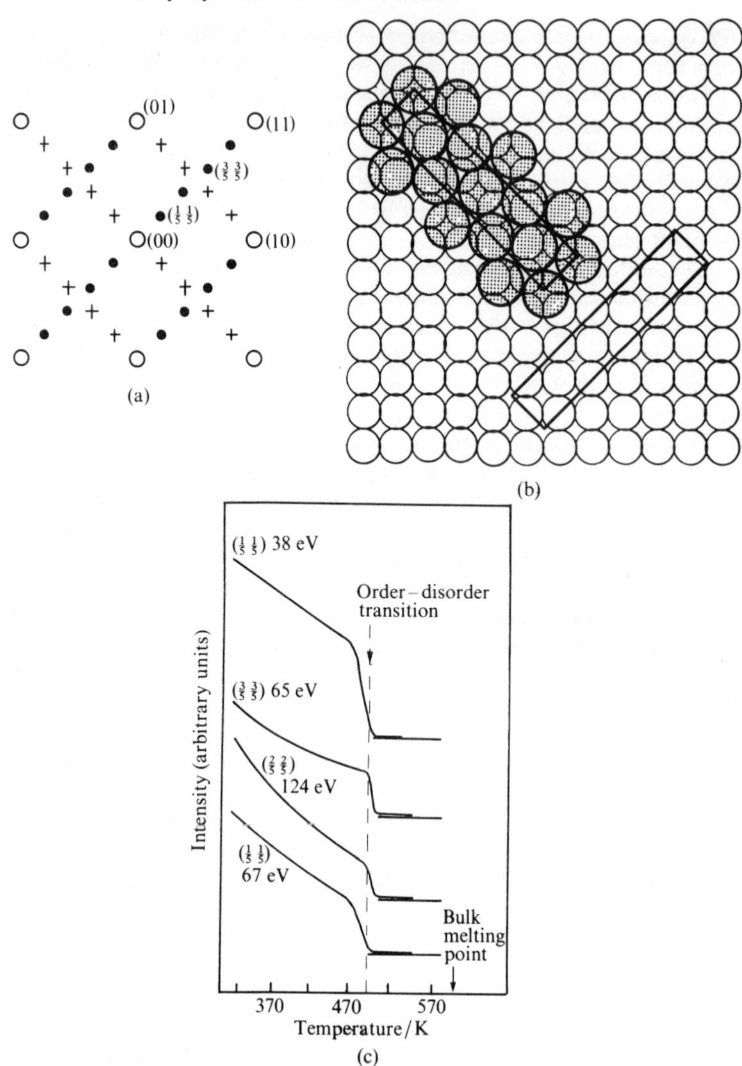

FIG. 5.5. (a) The Cu(100)c(5 x 1)–$R45°$ –Pb LEED pattern. ○ Cu substrate spots; ● spots due to one c(5 x 1) domain of lead; + spots due to the other domain of lead. (b) A possible model of a lead structure that would give the diffraction pattern of (a). The two domains with equivalent orientations with respect to the substrate (enantimorphs) are indicated. (c) Temperature-dependence of the LEED intensities of spots due to the lead overlayer. The transition is the sharpest observed in the copper–lead system. The curves are reversible with temperature. The zeros have been displaced for clarity of display. (After Henrion and Rhead 1972.)

metal upon a clean surface of a second metal. Some examples of such systems showing monolayer growth were mentioned in Chapter 2 because of their usefulness in calibrating electron spectroscopic techniques such as AES, XPS, and UPS. They can also be used to study melting processes by observing their LEED patterns and the drop in intensity of LEED spots with increasing temperature.

One system which appears to show a surface melting point below that of the bulk is monolayers of lead on various copper surfaces. This system was reported by Henrion and Rhead (1972). One example from their results is an ordered structure corresponding to a monolayer of lead in dense (111) – like packing upon the (100) surface of copper. From the LEED observations this pattern is called Cu(100)c(5x1) $R45^\circ$–Pb (Fig. 5.5(a) and (b)). All the spots in LEED the pattern can be accounted for by recognizing that there are two equivalent orientations of this lead structure upon the copper, as shown in Fig. 5.5(b). Upon heating this structure there is no change in the lead Auger signal at 93 eV up to 673 K, from which it can be concluded that there is little or no interdiffusion or alloying of the lead with the copper. However, the intensities of LEED spots due to the lead overlayer pass through a sharp drop which has a point of inflection at 498 K. The melting point of lead is 600 K. This drop is shown in Fig. 5.5(c).

Summary

Atomic vibrations in a solid surface can be measured using the temperature-dependence of the intensity of LEED spots in a way analogous to the Debye-Waller correction applied in X-ray diffraction experiments. The modes of vibration can be detected using a very high-energy resolution electron spectroscopy. It is found that the atomic vibrations have larger amplitudes at the surface than in the bulk, particularly in the direction of the surface normal. Also, surface vibrational modes known as surface phonons can be found and have energies of the order of a few tens of millielectronvolts. More extreme atomic motions occur when the atoms leave their equilibrium lattice sites and diffuse over the surface or assume a disordered array. Surface diffusion of single adatoms can be observed with FIM, and activation energies of the order of 1 eV are found to apply. Surface melting and surface phase changes can also occur and can be detected by LEED.

6. Surface properties: adsorption of atoms and molecules

A wide variety of events can occur when a molecule impinges upon a surface. It may be specularly reflected with no loss of energy or it may suffer a redistribution of momentum and be diffracted by the surface, again with no loss of energy. Alternatively, and more usually, it will lose energy to the atoms in the surface by exciting them vibrationally or electronically. If it loses only a small amount of its energy and does not become bound to the surface it may be inelastically reflected. On the other hand, it may lose sufficient energy to become effectively bound to the surface with a strength that will depend upon the particular kinds of atoms involved. If this occurs the molecule is said to be *accommodated*. by the surface – it has an energy appropriate to the temperature of the surface and has become *adsorbed*. As discussed in Chapter 5 it may diffuse about the surface until it picks up enough energy from thermal fluctuations to leave again or desorb. An ensemble of adsorbed molecules is called an *adlayer* and the average time of stay of a molecule upon the surface is called the *mean stay time*. Of course, even more complex events can also occur. For instance, an impinging molecule may dissociate before it can be adsorbed upon the substrate – a process known as dissociative adsorption. Adsorption of this kind is discussed at greater length by Bond (1974).

The experimental methods used to study these events can be microscopic or macroscopic. For example, a microscopic study might involve observations of the angular and energy distributions of atoms which had been scattered out of a monatomic and monochromatic incident beam. Such an experiment would be interpreted in terms of the interaction of individual atoms with the surface. A macroscopic example is the measurement of the density of the adlayer (*the coverage*) as a function of the pressure of the adsorbing species and of the surface temperature. This experiment might be interpreted in terms of an average binding energy for the whole surface, which may or may not be useful for yielding information on an atomic scale.

Adsorption is of practical significance in a wide variety of problems and processes. It is the first stage in the formation of an oriented overgrowth – *epitaxial* growth of thin films (p. 104). It is important in catalysis, where different adsorbates upon a catalytic surface may help or hinder the progress of a chemical reaction. It is important in vacuum technology where it is used for pumping gases out of chambers (e.g.

cyrogenic pumps) and is a nuisance if adsorbed gases have to be removed in order to improve the ambient pressure.

The wide range of motivations for studying adsorption, the availability of microscopic and macroscopic methods, and the large variety of surface events possible all conspire to make this area of surface studies into a broad and developing subject; only a few aspects can be touched upon here.

Adsorption processes

In the case of *physical adsorption* (or physisorption), an adsorbed molecule is bound to the surface via a rather weak Van der Waals type bond. This type of bond involves no charge transfer from the substrate to the adatom or vice versa. Rather, the attractive force is provided by the instantaneous dipole moments of the adatom and its nearest-neighbour surface atoms (see e.g. Kittel 1967). The interaction can be described by the potential-energy diagram shown in Fig. 6.1. An incoming

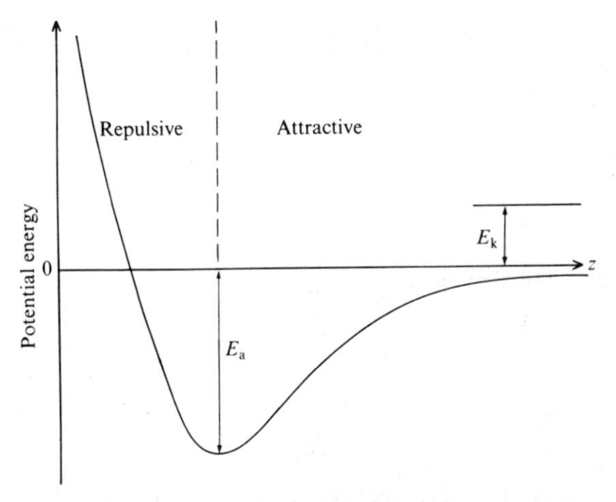

FIG. 6.1. Potential energy of an adatom in a physisorbed state on a planar surface as a function of its distance z from the surface.

molecule with kinetic energy E_k has to lose at least this amount of energy in order to stay on the surface. It loses energy (accommodates) by exciting lattice phonons in the substrate and the molecule then comes to equilibrium in a state of oscillation in the potential well of depth equal to the binding or adsorption energy E_A. The kinetic energy associated with these oscillations is that appropriate to the substrate temperature. In order to leave the surface the molecule must acquire enough energy to surmount the potential-energy barrier E_A. The desorption energy is thus equal to the adsorption energy.

The binding energies for physisorbed molecules are typically 0·25 eV or less. Such bonds are found in the adsorption of inert gases upon metals and glasses. If τ_0 is the period of a single-surface atom vibration in the well of depth E_A (Fig. 6.1) then the stay time τ of this atom upon the surface is given by

$$\tau = \tau_0 \exp(E_A/kT) . \tag{6.1}$$

The times τ_0 are usually of the order of 10^{-12} s and eqn (6.1) can be used to show that stay times greater than about 1 s will not occur until $E_A \gtrsim 28\ kT$. Thus, an adsorption energy of 0·25 eV will give $\tau > 1$ s only below temperatures of about 100 K. Eqn (6.1) is simply related to the frequency with which an atom can escape from a potential well described in eqn (5.3) (p. 84). The derivation of eqns (5.3) and (6.1) is given by Frenkel (1946).

It is more usual for electron exchange to occur between an adsorbed molecule and the surface, in which case a rather strong bond is created with the surface and the molecule is said to be *chemisorbed*. The most extreme case of chemisorption occurs when integral numbers of electrons leave the adsorbed molecule and stay on the nearest substrate atom (or vice versa). This would be a pure ionic bond. More usually there is an admixture of the wavefunctions of the valence electrons of the molecule with the valence electrons of the substrate into a new wavefunction. The electrons responsible for the bonding can then be thought of as moving in orbitals between substrate and adatoms and a *covalent* bond has been formed. A simple example of the potential energy diagram for chemisorption is shown in Fig. 6.2. Some of the impinging molecules are accommodated by the surface and become weakly bound in a physisorbed state (also called a precursor state) with binding energy E_p. During their stay time in this state electronic or vibrational processes can occur which allow them to surmount the small energy barrier E_c (activation energy $E_c + E_p$) and electron exchange occurs between the adsorbate and substrate. Each adatom now finds itself in a much deeper well E_A. It is chemisorbed. The range of binding energies (heats of adsorption) in chemisorption is quite large, extending from about 0·43 eV for nitrogen on nickel to about 8·4 eV for oxygen on tungsten.

The theoretical description of chemisorption is very complex and, as yet, far from complete (e.g. Schrieffer 1972). However, a clue as to one way in which an adatom may change from a physisorbed to a chemisorbed state can be obtained by thinking about the processes that can occur as an atom approaches a simple free-electron-like metal surface

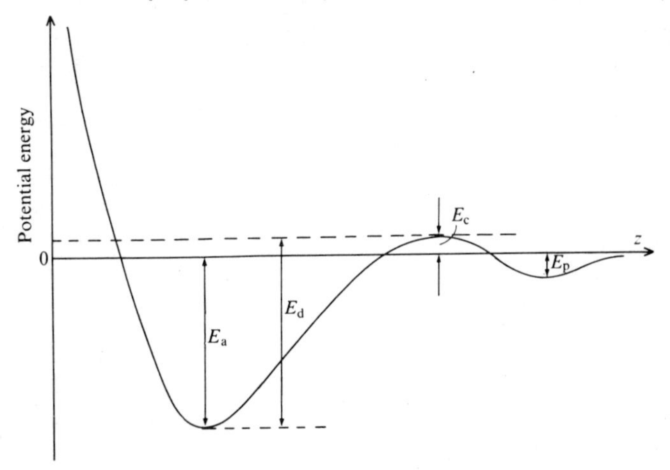

FIG. 6.2. A simple version of a potential energy diagram for chemisorption on a planar surface. Note that, once chemisorption has occurred, the desorption energy E_d is greater than the adsorption energy E_a. The potential wells contain discrete energy levels which correspond to the allowed vibrational states of the adatom. Strictly, two diagrams should be used, because the system changes once chemisorption has occurred.

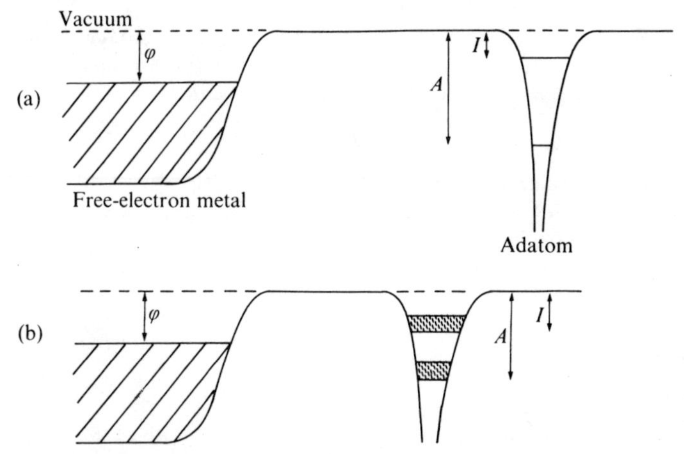

FIG. 6.3. (a) Electronic levels for an adatom so far from the surface of a free-electron metal that the levels of neither are disturbed. (b) The adatom is now closer to the metal and interacts weakly with it. The affinity and ionization levels become broadened into bands. The ionization potential I is defined as the energy required to remove an electron from an orbit and place it at rest, outside the atom. The electron affinity A is defined as the energy gained in taking an electron at rest just outside an atom and placing it in a vacant orbit.

(Fig. 6.3). If the ionization energy I of the adatom is less than the work function ϕ of the metal then an electron will be transferred to the metal and the adatom becomes ionized. An example of such a process is caesium (I = 3·87 eV) on tungsten ($\phi \simeq$ 4·5 eV). On the other hand, if the electron affinity A of the adatom is greater than ϕ an electron will be transferred from the substrate to the adatom. An example of this process might be a flourine atom (I = 3·6 eV) on caesium ($\phi \sim$ 1·8 eV) – the compound caesium fluoride is formed. The third case occurs if $I > \phi > A$, for here the atom is stable in its neutral state. Hydrogen, for instance, has I = 13·6 eV and A = 0·7 eV, and so may be expected to form neutral bonds with most metals (ϕ = 4–6 eV).

This kind of argument depends upon the notion that the electron levels of the adatom and the metal are not disturbed by each other. Of course, this is not strictly possible. When the adatom approaches the surface it first interacts weakly with the metal and electrons can tunnel between adatom and solid. This process will shift and broaden the adatom states (Fig. 6.3(b)) and both the affinity and ionization levels become narrow bands. The tails of both these bands may overlap the Fermi level of the metal and so both affinity and ionization states may be partially filled. As the adatom approaches more closely, entirely new electronic structure may appear as the strong interaction between solid and adatom modifies the electronic levels of both.

As more atoms or molecules arrive at the surface an adlayer begins to form. If the rate of arrival of particles is J then the coverage θ, measured in atoms per square centimetre, is related to the stay time τ by

$$\theta = \alpha_c J \tau, \tag{6.2}$$

where α_c is the *condensation coefficient*. It is simply the probablility that an impinging particle will be accommodated upon the surface. More easily measured is the sticking coefficient S, which is the rate of increase of coverage θ with total exposure M to impinging particles,

$$S = \partial\theta / \partial M, \tag{6.3}$$

where

$$M = \int J \mathrm{d}t. \tag{6.4}$$

Both S and θ can vary with M in ways which depend upon the details of the interaction between adatoms and substrate and upon the topology. Some examples of what can occur in practice are indicated in Fig. 6.4, where the coverage is expressed as a fraction of a monolayer instead of in atoms per square centimetre.

Because it increases coverage, the arrival of more adatoms reduces the distance between nearest-neighbour adatoms, and interactions between

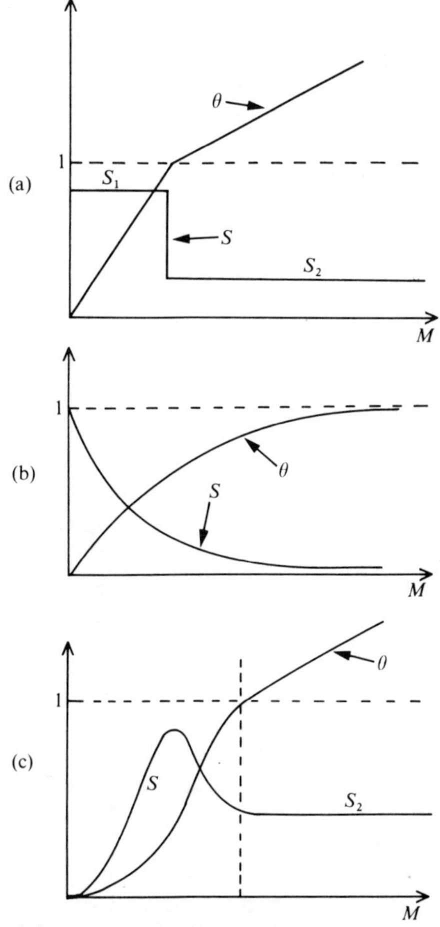

FIG. 6.4. Some plausible variation of sticking coefficient S and coverage θ with exposure M. (a) The adlayer forms on the substrate with constant sticking coefficient S_1 until a monolayer ($\theta = 1$) is formed. Then a new value of S – that of adsorbate growing upon adsorbate – is applicable. Such behaviour is seen for some metals condensing upon others (e.g. silver on nickel). (b) Incident atoms impinge directly upon unoccupied adsorption sites and fill them. As coverage increases the number of available sites falls and so the sticking coefficient falls. Finally, a monolayer is formed and no further impinging atoms stick. Such behaviour might be shown by gas adsorption on a metal. (c) Here the adsorption site is the perimeter of a nucleus of several adatoms in a cluster. As the clusters grow their perimeters increase in length and S increases. Then they begin to touch and coalesce and S decreases again. Finally the value of S is that pertaining to adatoms growing upon the adsorbate. Such behaviour can be exhibited by metals condensing upon alkali halides.

them become important. One effect of this interaction can be to order the adatoms crystallographically. Such ordered adsorption is usually an example of the beginning of epitaxial growth.

Experimental observations of chemisorption

The two most commonly used observables for characterizing chemisorption are the change in work function $\Delta\theta$ on adsorption and the heat of adsorption (or desorption). In order to obtain a more detailed understanding of the interaction of an adatom and a substrate it would be useful to be able to describe mathematically the shapes of the curves in Figs 6.1, 6.2, and 6.3 and to be able to place the electronic states. To obtain information on this adatom–solid force law is difficult, but feasible, as outlined in the following paragraphs.

1. Work-function changes. As described in Chapter 4, if an adatom donates an electron to the conduction band of the substrate the work function decreases on adsorption. Conversely, electron transfer to the adatom increases the work function. Thus the sign of $\Delta\theta$ immediately gives information about the direction of charge transfer. In many circumstances the adatoms are merely polarized by the surface attraction. In this case they may be thought of as being polarized normal to the surface. If the positive pole is at the interface (the negative pole on the vacuum side) the work function of the substrate will be increased and the opposite will occur if the negative pole is at the interface.

If the substrate is an insulator or a semiconductor then the density of surface states may be sufficiently large that it is these that will control the adsorption and not the underlying band structure.

2. The depth E_d of the wall in Fig. 6.2 can be measured by using the technique of *flash desorption*. The sample is heated up rapidly in a chamber of known volume V and the partial pressure of the adsorbate measured as a function of time with a sensitive mass spectrometer (e.g. Gomer 1967). As the temperature of the substrate passes through the energy corresponding to E_d a sharp burst of pressure due to desorption of the adatoms is observed. From measurements of the desorption rate as a function of temperature a value for E_d can be derived.

3. Another technique for obtaining values of E_d is to measure the mean stay time τ of an adatom as a function of substrate temperature T and then use eqn (6.1). Hudson (1967) has developed some elegant methods for making such a measurement (e.g. Sandejas and Hudson 1967) and one of his experimental arrangements is sketched in Fig. 6.5. By means of a shutter in the well collimated, monochromatic beam of incident atoms (cadmium in the case shown) the substrate can be exposed to a sudden flux of impinging atoms. Atoms leaving the surface after a

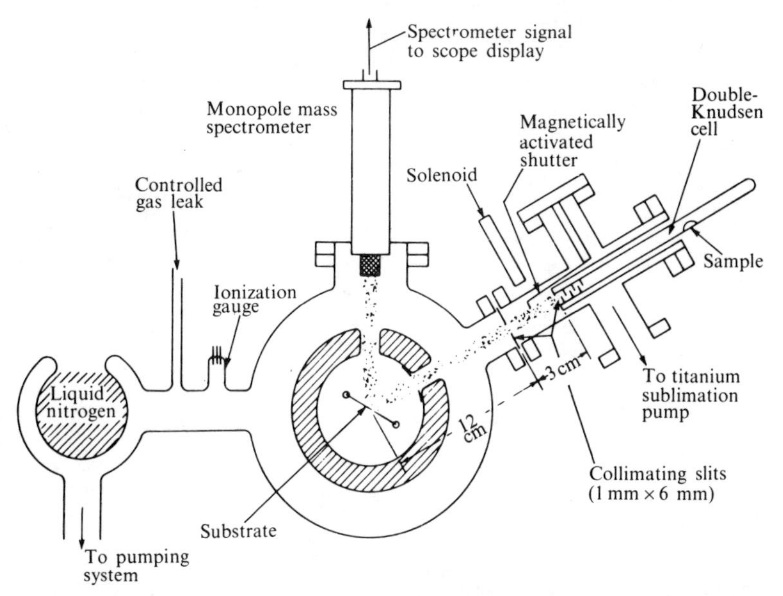

FIG. 6.5. A molecular beam experiment to study the adsorption of cadmium upon polycrystalline tungsten (Sandejas and Hudson 1967). The temperature of the sample of cadmium controls the flux of cadmium atoms falling upon a set of collimating slits. The temperature of the collimating slits controls the mean velocity of cadmium atoms reaching the substrate. The liquid-nitrogen dewar around the substrate ensures that atoms scattered from the substrate do not subsequently reach the detector unless they pass directly through the detector aperture. (From Sandejas and Hudson 1967.)

mean time τ are detected by a mass spectrometer. The spectrometer signal can be observed on an oscilloscope as the incident beam is periodically 'chopped' by the shutter. The value of τ is derived from the exponential rise and decay of the signal due to the desorbing atoms. An example of a plot of τ (on a logarithmic scale) versus $1/T$ for cadmium on clean polycrystalline tungsten is shown in Fig. 6.6. The slope of these plots gives the values of E_d indicated beside the lines. It appears that, in this case, the first monolayer or so of cadmium is bound tightly to the tungsten surface and subsequent cadmium atoms are bound more loosely. Detailed interpretation of the magnitude of E_d for each phase is often difficult, however.

4. Insight as to the energy levels of adatoms (Fig. 6.3(b)) has been obtained for hydrogen and deuterium on W(100) and W(110) by energy-analysing field-emitted electrons from these tungsten surfaces as a function of the coverage of adsorbate. The experimental arrangement is to combine a field-electron emitting tungsten tip (Chapter 4) and a high energy

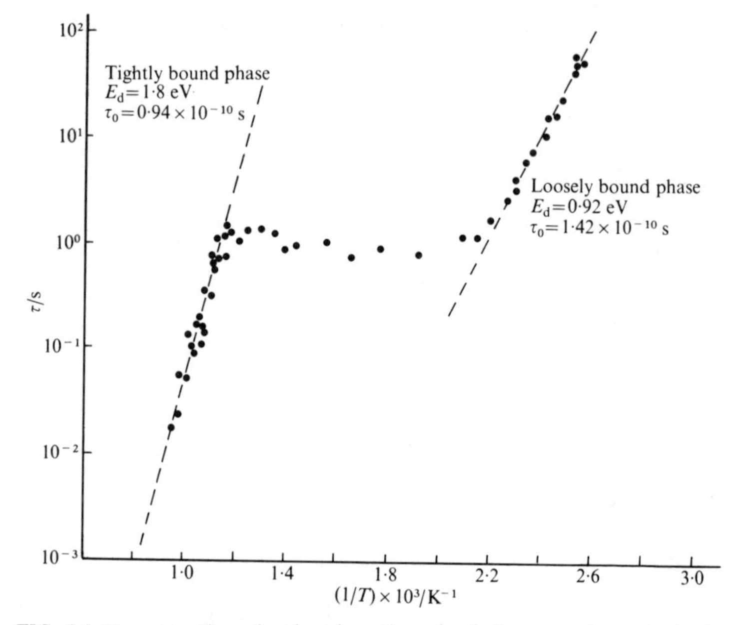

FIG. 6.6. Mean stay times for the adsorption of cadmium on polycrystalline tungsten. (From Sandejas and Hudson 1967.)

resolution electron energy analyser (the CHA of Fig. 2.3(d), p. 16). This sophisticated combination has been used by Plummer and Bell (1972). If there are empty states in a narrow band of the adsorbate-substrate combination (e.g. *A* or *I* in Fig. 6.3(b)) then electrons can tunnel from the substrate through the adsorbate without energy loss. This process is called *elastic tunnelling resonance*. On the other hand, field-emitted electrons may excite electronic or vibrational states of the adsorbate-substrate complex and, in so doing, lose energy. Such a process is called *inelastic tunnelling*. By comparing the energy distributions of field-emitted electrons from the clean tungsten surface with those at various coverages of adsorbate these two processes can be identified and something learned about the electronic states of the adsorbate. An example due to a single barium adatom upon W(111) plane is shown in Fig. 6.7. At the present time the technique is too new for firm conclusions to be drawn, but a combination of the results obtained in this way with flash adsorption and work-function measurements should be a very powerful means of understanding the character of the adatom-substrate bond.

In addition to the information yielded by all the methods described above it is important to realize that not only do the adatoms interact

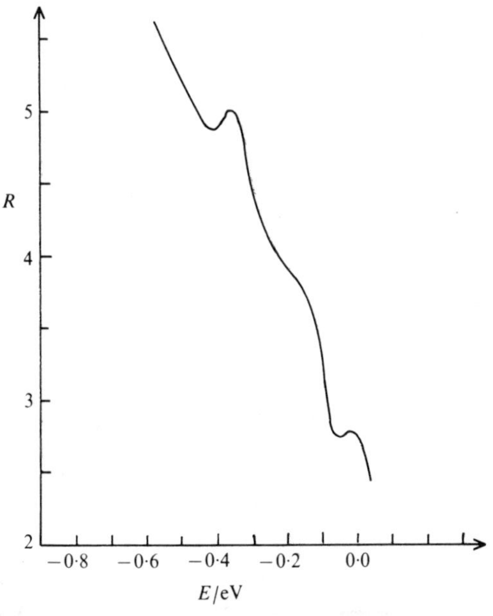

FIG. 6.7. The energy distribution of field-emitted electrons passing through a single barium atom upon a W(111) face. The vertical scale is the enhancement factor *R* which is the ratio between the energy distribution after adsorption and the energy distribution of electrons emitted from the clean W(111) surface. (After Plummer and Young 1970.)

with the substrate atoms but they also interact amongst themselves. If the substrate is a single-crystal surface then it is very common for atomically ordered adlayers to form. The unit mesh (Chapter 3) of such an adlayer is determined by the interplay of the bonding with the substrate and the bonding between the atoms. These ordered adlayers have been observed using LEED in hundreds of different adlayer–substrate combinations and some are tabulated by Somorjai (1972) and by Prutton (1971). There is, at present time, no theory which enables a prediction to be made of the size and shape of the unit mesh of adatoms that will be observed for a chosen adlayer–substrate combination. However, amongst the observations some patterns can be seen which give clues as to tendencies. Somorjai (1972) points out that adsorbed atoms or molecules of monolayer thickness tend to obey three 'rules':

1. They tend to form surface structures in which the adatoms are close packed. Thus, they grow with the smallest unit mesh permitted by the dimensions of each adatom, the adatom–adatom, and the

adatom–substrate interactions. An example of this effect is shown in Fig. 6.8 which is for Ni(100)c2–O – a more closely packed arrangement is not possible because of the large size of the adsorbed oxygen ions.

2. They tend to form ordered structures with the same rotational symmetry as the substrate.

3. They tend to form ordered structures with unit mesh sizes rather simply related to the substrate mesh size. Thus (1×1), (2×2), $c(2 \times 2)$, or $(\sqrt{3} \times \sqrt{3})-R30^\circ$ are all commonly observed. It is useful to draw out this last mesh using the information given in Chapter 3 because it is a particularly simple arrangement of the adatoms with respect to the substrate atoms.

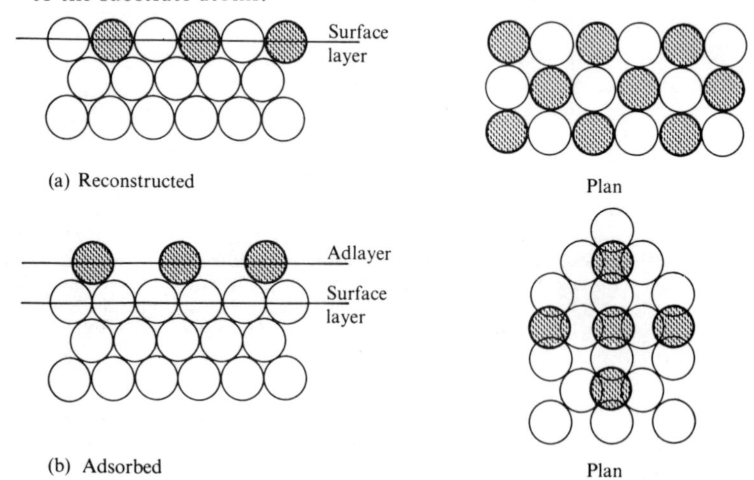

FIG. 6.8. Two possible surface structures to explain Ni(100)c(2 × 2)–O. Oxygen atoms (or ions) – open circles; nickel atoms (or ions) – hatched circles. (a) Reconstructed surface. Each oxygen and each nickel ion in the surface need not have the same charge provided that the whole layer is charge neutral. (b) Classical adlayer with no reconstruction.

A major difficulty associated with trying to understand and apply these 'rules' arises when an attempt is made to decide whether or not the adatoms are situated in a separate layer on top of the substrate surface. If they are then the 'rules' can be understood in terms of rigid-ball models of adlayer and substrate surface. If they are not then presumably the adatoms are incorporated into the surface layer, place-exchange of substrate atoms and adatoms has occurred and the surface is said to be reconstructed. Reconstruction must be the beginning of the formation of a compound of the adatoms and the substrate atoms and for this to

occur charge exchange is necessary between the atoms. Under such circumstances billiard-ball models may be dangerously misleading.
Two examples of chemisorption will be discussed below with a view to illustrating both the points made above and the techniques described in earlier chapters.

Ni(100)–O

The bulk oxide of Ni–NiO is a refractory material that melts at 2200 K. This suggests that the oxygen–nickel bond is very strong – a suspicion borne out by the value of about 2 eV for the desorption of oxygen from an ordered layer on Ni(100) (Brennan and Graham 1966). Also, NiO has the NaCl crystal structure and is a very ionic material, from which a simple view might be that it is made up of Ni^{2+} and O^{2-} ions. Thus, it might be anticipated that as oxygen molecules arrive at a Ni(100) surface they would be dissociated to oxygen atoms, ionized to form O^{2-} ions, and so bound by ionic attraction. As the coverage of O^{2-} ions increases the Coulomb repulsion between O^{2-} ions increases until a point will be reached where a charge neutral layer consisting of an equal number of Ni^{2+} and O^{2-} ions will have a lower energy than a layer of only O^{2-} ions. Unfortunately, such simple-minded arguments are immediately confounded by the fact that it costs energy (6·8 eV) to produce O^{2-} from O as well as costing energy to lift the two electrons from the top of the conduction band of Ni to form Ni^{2+} (the work function of Ni(100) is about 4 eV). This difficulty is resolved by including the fact that a lattice of ions creates an electrostatic crystal field which must be included in evaluating the energy of the system. An O^{2-} ion may not be stable when isolated but can easily be stable if surrounded by attractive Ni^{2+} ions. A careful set of considerations of this style is given by Carroll and May (1972) who conclude that reconstructed adlayers containing nickel and oxygen ions will be energetically favourable but that the oxygen ions need not necessarily carry exactly two electronic charges each.

LEED observations indicate that two stable phases exist for oxygen on Ni(100). At a coverage of a quarter monolayer there is a Ni(100)-p(2 x 2)–O LEED pattern and at a coverage of a half monolayer there is a Ni(100)c(2 x 2)–O LEED pattern. The latter is of particular interest in that similar diffraction patterns are observed for other adsorbates and substrates (e.g. Ni(100)c(2 x 2)–S, Cu(100)c(2 x 2)–O). The question of great interest is: are the oxygen atoms in this half-monolayer phase adsorbed on the surface *or* are they incorporated into the substrate by reconstruction of the surface layer? Two possible models for these alternatives are shown in Fig. 6.8. One way of choosing between these models is to carry out a calculation of LEED I(V) curves (as described in Chapter 3 for each model and compare these theoretical curves with experimental I (V) determinations. This has been done (see Pendry 1974), and the conclusion at present is that such comparisons suggest that a classical adlayer is formed (Fig. 6.8(b)). At the same time, considerations of heats of adsorption and Auger electron spectra (see the discussion in May and Carroll (1972)) suggest a reconstructed surface like Fig. 6.8(a) with mixed ionic and covalent bonding. This is supported by RHEED observations due to Garmon and Lawless (1970), who conclude that the Ni(100)c(2 x 2)–O pattern is due to the formation of an oxide of composition Ni_3O. The situation remains controversial. As the controversy depends upon the interpretation of two different models (a model for LEED I(V) curves and a model for the oxidation process) it may not be resolved until a third kind of observation, less dependent on models, can be made on the system.

Pd(111)–CO

The work of Ertl and Koch (1972) provides a good example of the way in which several techniques can be combined so as to show conclusions about the atomic details of an adsorption process. The adsorption of carbon monoxide upon the clean Pd(111) surface was studied by means of LEED, Auger spectroscopy, mass spectroscopy, work-function measurement, and flash desorption. Flash desorption showed a single peak at about 470 K, from which it is concluded that there is a single binding state on Pd(111). The area under this peak could be used to determine the coverage of carbon monoxide and so the work-function change on adsorption could be related to coverage. The work function increased on adsorption which suggests polarization of the carbon monoxide admolecules with a negative pole towards the vacuum. The variation of work function with coverage, taken together with a knowledge of the arrival rate of carbon monoxide molecules impinging from the ambient carbon monoxide atmosphere, enables the variation of sticking coefficient S with coverage θ to be evaluated. This shows initial behaviour rather like Fig. 6.4(a) with $S_1 = 1$ until θ is about 0·2 monolayers. S then falls linearly to zero at $\theta = 0{\cdot}5$ monolayers. By measuring the partial pressures of carbon monoxide and the work function at various temperatures the isosteric heat of adsorption could be estimated (isosteric means constant θ) and is found to be 1·47 eV up to $\frac{1}{3}$ of a monolayer.

The first ordered LEED pattern at room temperature is Pd(111)-$(\sqrt{3} \times \sqrt{3})R30°$–CO (Fig. 6.9(a)) which occurs at $\theta = \frac{1}{3}$. There are three equivalent domain orientations of this structure which are indicated in Ertl and Koch's proposed model of the carbon monoxide adsorption

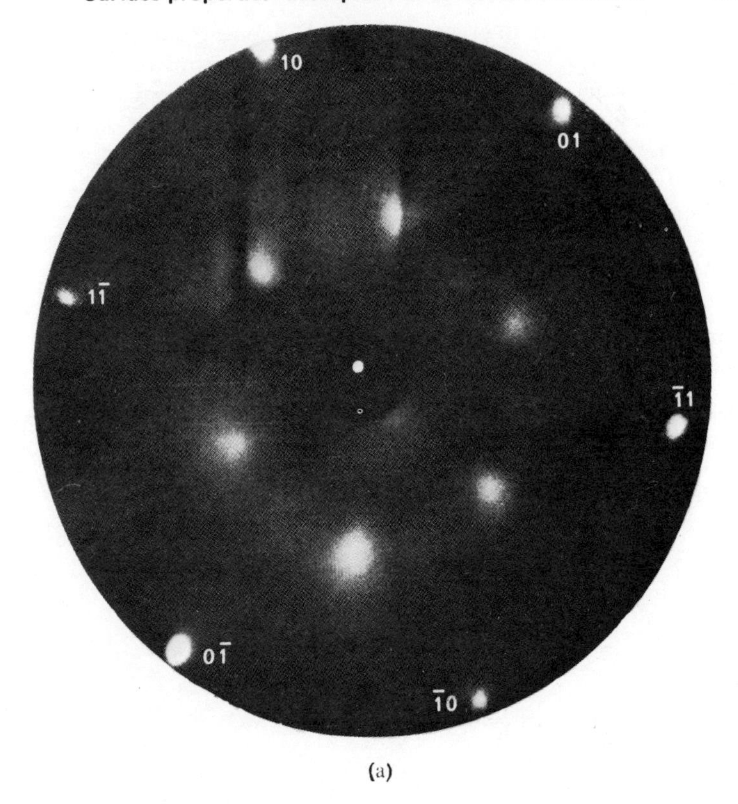

(a)

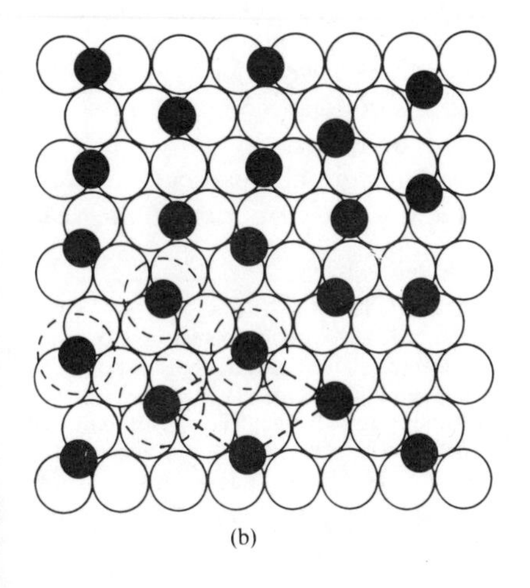

(b)

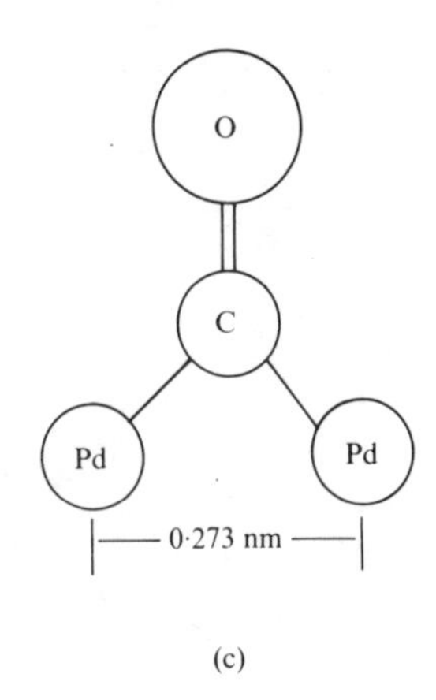

(c)

shown in Fig. 6.9(b). In order to suggest this model the work function changes described above, together with infrared spectroscopic studies on the same system by other observers, are used to conclude that the carbon is adjacent to the metal surface in a 'bridge-bonded' site indicated in Fig. 6.9(c). The axis of the carbon monoxide molecule is normal to the surface. The size of the carbon monoxide molecule is such that there is space only for one in each unit mesh of the adlayer. Similar structure models for carbon monoxide adsorption on Pd(100), Rh(111), Ni(100), and Ir(111) have been proposed by other authors.

Epitaxial processes

The gas–solid interactions described above may be the first stages in the growth of an oriented single-crystal film of one material upon a single-crystal substrate of another. This process is called epitaxial growth. An example is the formation of an oriented layer of nickel monoxide upon nickel – an epitaxial process that can proceed in the sequences:

$$\text{Ni(100)} + \text{O}_2 \xrightarrow{\text{R.T.}} \underset{\text{adlayer}}{\text{Ni(100)p(1} \times \text{1)–O}} + \underset{\text{substrate}}{\text{Ni(100)}}$$

$$\text{Ni(100)p(1} \times \text{1)} + \text{O}_2 \xrightarrow{\text{R.T.}} \underset{\text{adlayer}}{\text{Ni(100)c(2} \times \text{2)–O}} + \underset{\text{substrate}}{\text{Ni(100)}}$$

$$\text{Ni(100)c(2} \times \text{2)–O} + \text{O}_2 \xrightarrow{400\,^\circ\text{C}} \underset{\text{epitaxial film}}{\text{NiO(100)}} + \underset{\text{substrate}}{\text{Ni(100)}}$$

Examples of epitaxial processes are found not only in such gas-solid interactions but also in the deposition of many materials upon single crystals of very many others. The subject is reviewed by Chopra (1969).

Just as the situation with respect to adsorption is complex so it is with respect to epitaxial growth. There is no theoretical fabric which enables predictions to be made as to whether or not a particular material will grow epitaxially upon another, and, if so, in what orientation. What

FIG. 6.9. (a) The LEED pattern at 34 eV for Pd(111)–CO. It is a $\sqrt{3}\,R30°$ –CO pattern. Room temperature; $\theta = \frac{1}{3}$. (b) Ertl and Koch's model to account for this LEED pattern. The open circles are the palladium surface atoms, the solid discs are carbon atoms, and the broken circles are oxygen atoms. One of the three equivalent domain orientations is indicated by its unit mesh. (c) The way the carbon monoxide molecule is thought to 'stand up' on the surface with a bridge bond between palladium atoms to the carbon. The carbon monoxide is thus bonding with two-fold coordination unlike the four-fold coordination of the oxygen in Ni(100)c(2 x 2)–O shown in Fig. 6.7.

process occurs depends at least upon the adatom–adatom, substrate-atom–substrate-atom, and substrate-atom–adatom-bond strengths, the incident flux, the substrate temperature, and the surface diffusion coefficient of the adatom. Very often insufficient data are available about all these parameters for the observations of epitaxial processes to be arranged in meaningful patterns.

Most studies of epitaxial processes have used the techniques of electron microscopy (Pashley 1970) which normally involves removing the epitaxial film from its substrate and transferring it, through the air, from the deposition system to the electron microscope. Recently, *in situ* observations in special UHV electron microscopes have been possible and, in addition, the techniques of LEED and Auger electron spectroscopy have been used. The impact of these surface techniques upon the understanding of epitaxial processes is reviewed by Bauer and Poppa (1972).

One way of classifying processes of epitaxial growth is by the mode of growth, as indicated in Fig. 6.9. After specifying the mode of growth it is necessary also to define the orientation of the deposited layer with respect to the substrate. If the adlayer is only of the order of a monolayer thick then the notation used so far in this book is adequate to describe the deposit orientation. Thicker oriented films are described by specifying the deposit plane which is parallel to the substrate surface plane and also a direction in the deposit surface plane which is parallel to a direction in the substrate surface plane. Examples are given later.

Provided that the deposit does not alloy with the substrate, or that no gross changes in the substrate surface structure occur during deposition (e.g. dissociation of a substrate due to electron bombardment or chemical reaction between substrate and deposit) then the modes of Fig. 6.10 can be understood in terms of the relative *surface energies* of the deposit and substrate materials. The surface energy is simply the excess internal energy of the solid–vacuum system over that of an imaginary system with two homogeneous phases separated by an ideally discontinuous change at a mathematical surface between them (e.g. Blakely 1973). The surface energy is different from bulk energy because of the broken bonds at the surface and the possible relaxation or rumpling (Chapter 1). For nucleation to occur (Fig. 6.10(a)) the surface energy of the deposit material is high compared to that of the substrate. For monolayers to form at the substrate surface (Fig. 6.10(b) or (c)) the deposit must have a lower surface energy than the substrate. Bauer and Poppa (1972) distinguish monolayer growth (Fig. 6.10(b)), from nucleation after monolayer growth (Fig. 6.10(c)) by recognizing that if the deposit is strained so as to 'match' deposit lattice spacings to substrate lattice spacings then the growth mode will depend upon the relative sizes of the deposit strain energy and the deposit surface energy. If the strain energy in the deposit is low compared to its surface energy then monolayer growth is expected. If, however, the deposit strain energy is high it may become defective in some way after a monolayer is formed. This defectiveness may show up as dislocations in an otherwise flat layer or the material may nucleate on top of the first monolayer as in Fig. 6.10(c).

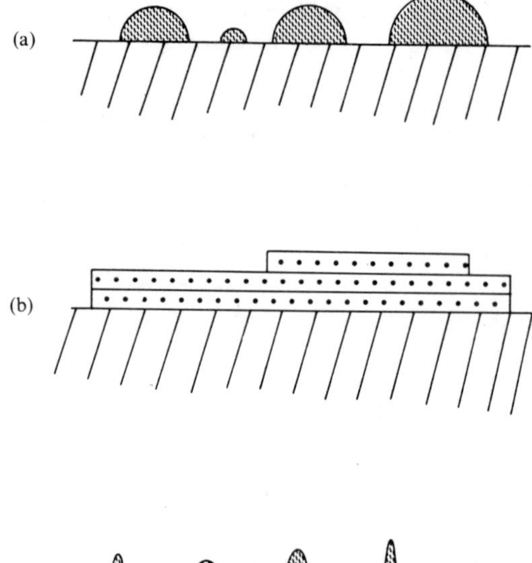

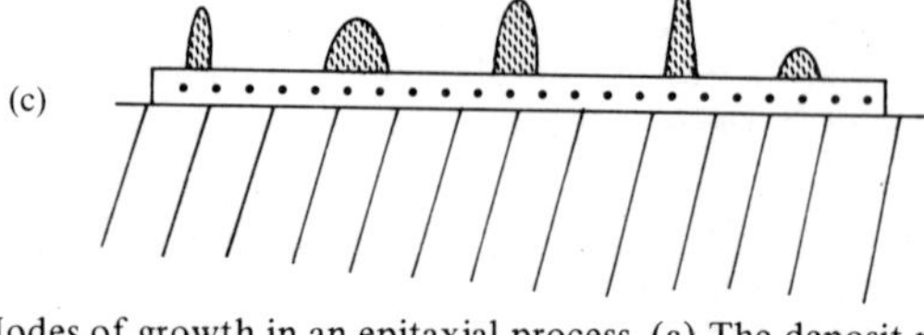

FIG. 6.10. Modes of growth in an epitaxial process. (a) The deposit nucleates on the substrate surface either randomly, as adatoms meet by accident and form stable clusters, or at special defect or impurity sites in the substrate surface (see Fig. 5.4.) Nuclei then grow by addition of adatoms directly from the impinging vapour or from surface diffusion. The nuclei may rotate on the substrate. Coalescence to a film occurs later and orientation changes can happen at this stage of growth. (b) The deposit grows in monatomic adlayers first upon the substrate and subsequently upon itself. (c) The first atoms arriving at the substrate form an atomic monolayer and subsequent atoms nucleate to form islands on top of the monolayer.

In addition to these different growth modes, different relative orientations of deposit and substrate can occur. If the deposit grows in parallel orientation with the substrate surface, the deposit atoms building up as if to continue the atomic structure of the substrate, then the process is said to be *pseudomorphic*. If a film of metal is grown upon a single-crystal surface of another metal which has a lattice spacing close to that of the deposit, then pseudomorphic epitaxy is often observed. An example of such a system is nickel on copper. Nickel and copper differ in bulk lattice spacings by only 2·5 per cent and nickel is found to grow pseudomorphically upon Cu(100) and Cu(111) surfaces with the mode of Fig. 6.10(b). As the thickness of nickel is increased it stores more and more strain energy until a critical thickness is reached where dislocations can be introduced and the nickel can relax towards its bulk lattice constant. This case is discussed, for instance, by Joyner and Somorjai (1973). If strict pseudomorphism were occurring here the depositing nickel atoms would be located in copper 'sites' as indicated in an exaggerated way in Fig. 6.11. It should be possible to use LEED I(V) studies (Chapter 3)

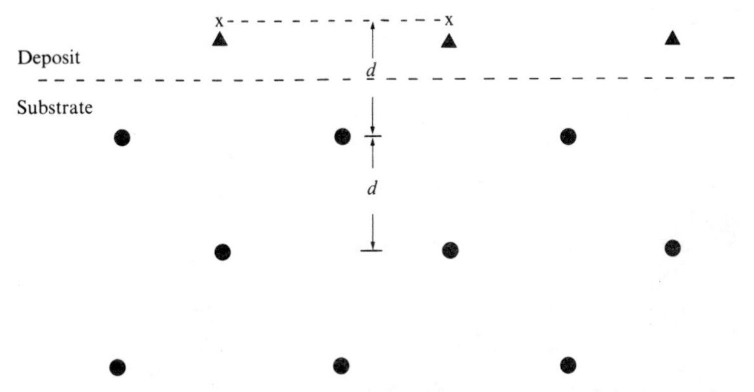

FIG. 6.11. The pseudomorphic growth problem. In strict pseudomorphism depositing atoms, e.g. nickel would be located at substrate atom sites marked X (e.g. copper). If the nickel–copper spacing d is not strictly pseudomorphic the deposit atoms may be located with a nickel–nickel spacing or some intermediate spacing (marked ▲) appropriate perhaps to an interfacial alloy.

to determine the vertical spacing of the nickel adatoms, but this problem has not yet been attempted. A similar example which has been studied by almost every technique described in this book is W(110)–Ag. Here, instead of strict pseudomorphism, the first monolayer is an Ag(111) layer which is strained to fit the tungsten atomic spacings along the W[001] direction. As more silver is deposited further monolayers grow but with different periodicities until finally an epitaxial silver layer is formed with Ag$\{111\}$ [112] parallel to W(110) $[1\bar{1}2]$. This case is described by Bauer and Poppa (1972).

On the other hand, a deposit may grow as a single crystal with the bulk lattice spacing from the earliest observable stages. Such a growth process is observed in the much studied alkali-halide–metal systems.

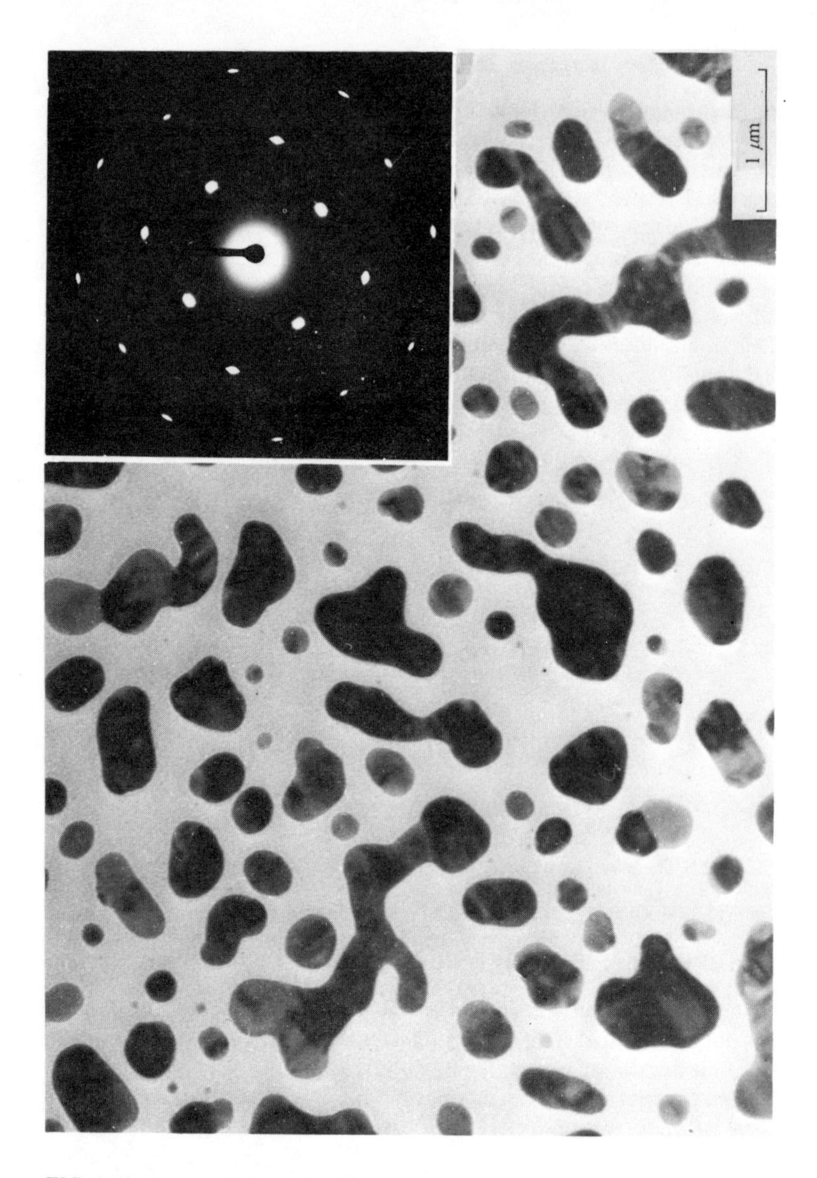

FIG. 6.12. A transmission electron micrograph of nominally 4 nm of silver deposited onto KC1(100) at 320 K. The silver islands are supported on an amorphous film of carbon about 20 nm thick added after the epitaxial deposition. The potassium chloride substrate was electron-bombarded during deposition – a process which improves the (100) epitaxy of the deposit. *Inset*. Transmission diffraction pattern of the same region as the micrograph. Obtained with electrons of energy 100 keV. (From Lord and Prutton 1974.)

These systems were of great interest because the epitaxial film could be dissolved off its alkali-halide substrate and floated onto a metallic support grid for study by transmission electron microscopy. All these systems nucleate as in Fig. 6.10(a) with islands usually in a (100) parallel epitaxial orientation or with a (111) plane parallel to the (100) cleavage face of the alkali-halide. An example of the parallel orientation in KC1(100)–Ag is shown in Fig. 6.12. In these cases the mismatch between the lattice constants of substrate and deposit can be very large (sometimes as much as 30 per cent) and so the strain energy that would be associated with pseudomorphic growth renders it unsupportable. It appears that (100) oriented islands are formed by clusters of four adatoms strongly bonded together and less strongly bound to the substrate. The orientation of this cluster is determined by its interaction with the substrate surface – as indicated in Fig. 6.13. In the case of metals upon alkali halides the anisotropic character of this interaction (which must be the feature determining the orientation of the cluster) is not very strong and not only can several orientations occur, but the balance

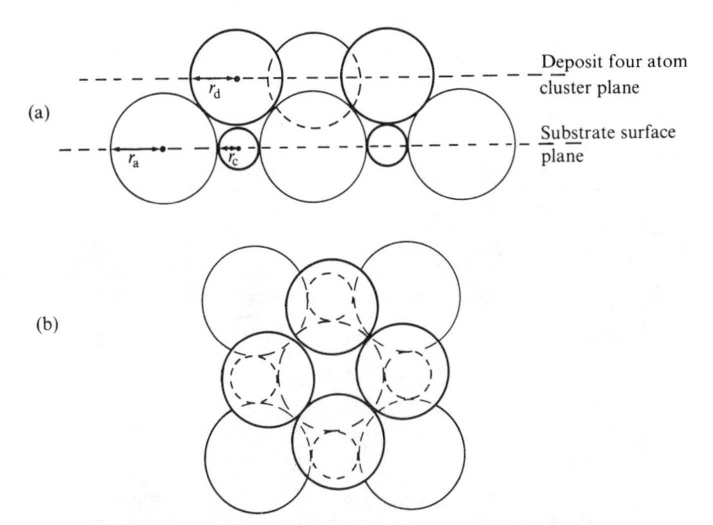

FIG. 6.13. The four-atom cluster for epitaxial nucleation with (100) [001] deposit parallel to (100) [001] substrate. (Due to Vermaak and Henning 1970.) (a) Section containing [001] in substrate and deposit; (b) plan looking down on a cluster in the (100) plane. Such clusters are possible if

$$r_{\mathrm{d}} < \frac{r_{\mathrm{a}}^2 - r_{\mathrm{c}}^2 + d^2}{(2r_{\mathrm{c}} + \sqrt{2}d - r_{\mathrm{a}})},$$

$$d = r_{\mathrm{a}} + r_{\mathrm{c}}.$$

between them can be disturbed by the process of coalescence. The orientation of the first continuous film is then not very clearly related to the dominating orientation amongst the initial clusters. The subject is complex and the interested reader is referred to recent reviews by Bauer and Poppa (1972) and Lord and Prutton (1974).

In the example of a four-atom cluster taking up a particular orientation upon an alkali halide (100) surface two important parameters in the theoretical description of the problem will be the adatom–substrate interaction potential and the adatom–adatom interaction potential. Little is known of the former and very little of the latter in spite of its importance in determining whether or not ordered overlayers will form. The experimental difficulty in observing such interactions is considerable but has been possible using the atomic resolving power of FIM (Chapter 3). If the adatoms are sufficiently strongly bound to the surface of the tip in a field ion microscope and yet still mobile enough to diffuse around and come into positions of minimum energy with respect to each other then adatom–adatom interactions can be observed. In some particularly elegant work Bassett (1973) has been able to observe stable two-atom clusters – dimers – of Ta_2, W_2, Ir_2, Pt_2, and WRe upon W(110) surfaces. Iridium is particularly striking in that it can form long parallel adatom chains (Fig. 6.14) which form about 0·15 nm apart.

If the adatom–substrate interaction is strong but the adatom–adatom

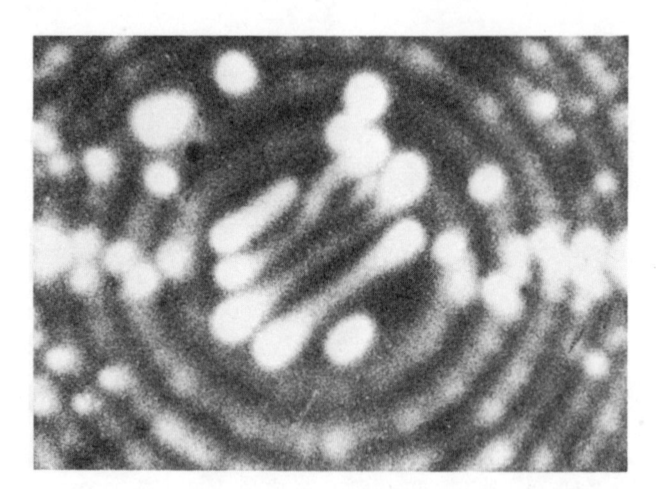

FIG. 6.14. Parallel adatom chains with a separation of 0·15 nm in a deposit of (80 ± 5) iridium atoms formed on a W(110) plane on a tip in an FIM. Iridium vapour was deposited at 78 K and then the tip was heated to 380 K. (After Bassett 1973.)

interaction is weak, then the growth mechanism of Fig. 6.10(c) – monolayer followed by nucleation – can occur. Such a mechanism is observed for alkali metals deposited upon tungsten (Mayer 1971).

The wealth of examples of epitaxial growth is as great as the paucity of explanations for their occurrence. It is to be hoped that clearer patterns will emerge as the great variety of parameters required to describe these processes are measured using the kind of techniques described in the chapters above.

Summary

The problems described in this chapter are amongst the most challenging and the least resolved in surface physics. The adsorption of single atoms upon single-crystal surfaces is the first step in many kinds of processes and yet it is poorly understood. It requires knowledge of the way electron exchange can occur with substrate, to what extent the substrate atoms relax about the adatom, and a precise knowledge of the position in which the adatom is located. Measurements of work function and energy distribution of emitted electrons and observations using LEED, Auger electron emission, UPS, XPS, and RHEED can help to provide this information but the power of combinations of these techniques has rarely been applied to one carefully characterized system. The success attendant upon this approach, in the few cases in which it has been used, encourages further work. The case of carbon monoxide adsorption on f.c.c. metal faces is an example of such an approach.

The same conclusions can be drawn about the second difficult problem described in this chapter – epitaxial growth. Again, a good example is W(110)–Ag. An understanding of the interplay of the important parameters – the adatom–substrate and the adatom–adatom interaction potentials – can be provided by combinations of the kinds of techniques described here but it is unlikely to be revealed by only one of them.

References

Chapter 1

Bond, G. C. (1974). *Heterogeneous catalysis: principles and applications* (OCS 18). Clarendon Press, Oxford.
Redhead, P. A. (1968). *The physical basis of ultra-high vacua.* Chapman and Hall, London.
Rosenberg, H. M. (1975). *The solid state.*(OPS 9). Clarendon Press, Oxford.
Yarwood, J. (1967). *High vacuum technique.* Chapman and Hall, London.

Chapter 2

Bassett, P. J., Gallon, T. E., Matthew, J. A. D., and Prutton, M. (1972). *Surf. Sci.* **33**, 213.
Bearden, J. A. and Burr, A. F. (1967). *Rev. mod. Phys.* **39**, 125.
Benninghoven, A. and Loebach, E. (1971). *Rev. Scient. Instrum.* **42**, 49.
Bishop, H. E. and Rivière, J. C. (1969). *J. appl. Phys.* **40**, 1740.
Fadley, C. S. and Shirley, D. A. (1970). *J. Res. natn. Bur. Stand.* **74**A, 543.
Gallon, T. E. and Matthew, J. A. D. (1972). *Rev. Phys. Technol.* **3**, 31.
—— Prutton, M., and Wray, L. (1971). *J. Vac. Sci. Technol.* **9**, 911.
Jackson, D. J., Chambers, A., and Gallon, T. E. (1973). *Surf. Sc.* **36**, 381.
Kittel, C. (1967). *Introduction to solid state physics.* Wiley, New York.
Kuhn, H. G. (1969). *Atomic spectra.* Longmans Green. London.
Maissel, L. I. and Chang, R. (1970). *Handbook of thin film technology.* McGraw-Hill, New York.
Müller, E. W. and Tsong, T. T. (1969). *Field ion microscopy.* Elsevier, New York.
Redhead, P. A. (1968). *The physical basis of ultra-high vacua.* Chapman and Hall, London.
Sevier, K. D. (1972). *Low energy electron spectrometry.* Wiley–interscience, New York.
Siegbahn, K., Nordling, C., Fahlman, A., Nordberg, R., Hamrin, K., Hedman, J., Johannson, G., Belgmark, T., Karlsson, S. E., Lindgren, I., and Lindberg, B. (1967). ESCA – *Atomic, molecular and solid state structure studied by means of electron spectroscopy.* Almquist and Wiksell, Uppsala.
Spicer, W. E. (1970). *J. Res. natn. Bur. Stand.* **74**A, 397.

Chapter 3

Davisson, C. J. and Germer, L. H. (1927). *Phys. Rev.* **30**, 705.
Estrup, P. J. and McRea, E. G. (1971). *Surf. Sci.* **25**, 1.
Gervais, A., Stern, R. M., and Menes, M. (1968). *Acta Cryst.* **24**, 191.
Heidenreich, R. D. (1964). *Fundamentals of transmission electron microscopy.* Wiley, New York.
Müller, E. W. (1951). *Z. Phys.* **131**, 136.
—— (1965). *Science* **149**, 591.
—— (1970). *Modern diffraction and imaging techniques in materials science* (eds S. Amelinckx, R. Gevers, G. Remault, and J. Van Landuyt). North Holland, Amsterdam.
—— Panitz, J. A., and McLane, S. B. (1968). *Rev. Scient. Instrum.* **39**, 83.

Pendry, J. B. (1974). *Low energy electron diffraction.* Wiley–Interscience, New York.
Prutton, M. (1971). *Metall. Rev.* **152**, 57.
Rosenberg, H. M. (1974). *The solid state* (OPS 9). Clarendon Press, Oxford.
Wood, E. A. (1964). *J. appl. Phys.* **35**, 1306.
Woolfson, M. M. (1971). *An introduction to X-ray crystallography.* Pergamon Press, Oxford.
Wormald, J. (1973). *Diffraction methods* (OCS 10). Clarendon Press, Oxford.

Chapter 4

Adams, D. L. and Germer, L. H. (1971). *Surf. Sci.* **27**, 21.
Beshara, N. M., Buckman, A. B., and Hall, A. C. (1969). *Proceedings of the Symposium on recent developments in ellipsometry.* North Holland, Amsterdam.
Fowler, R. (1933). *Phys. Rev.* **38**, 45.
Gadzuk, J. W. (1972). *J. Vac. Sci. Technol.* **9**, 591.
Heavens, O. S. (1964). Measurement of optical constants of thin films, in *Physics of thin films*, (eds G. Haas and R. E. Thun), Vol. 2. Academic Press, New York.
Kittel, C. (1967). *Introduction to solid state physics.* Wiley, New York.
Lunsford, J. H. (1972). *Adv. Catalysis* **32**, 265.
Malus, E. L. (1808). *Nov. Bull. Soc. philomath.* **1**, 266.
McKelvey, J. P. (1966). *Solid state and semiconductor physics.* Harper and Row, New York.
Müller, E. W. (1970). *Modern diffraction and imaging techniques in materials science* (eds S. Amelinckx, R. Gevers, G. Remault, and J. Van Landuyt). North Holland, Amsterdam.
Porteus, J. O. and Faith, W. N. (1973). *Phys. Rev.* **B8**, 491.
Rosenberg, H. M. (1974). *The solid state* (OPS 9). Clarendon Press, Oxford.
Rowe, J. E. and Ibach, H. (1973). *Phys. Rev. Lett.* **31**, 102.
Somorjai, G. A. (1972). *Principles of surface chemistry*. Prentice-Hall, Englewood Cliffs, New Jersey.
Tompkins, F. C. (1967). In *Gas–surface interactions* (ed E. A. Flood). Dekker, New York.
Vrakking, J. J., and Meyer, F. (1971). *Appl. Phys. Lett.* **18**, 226.
Wert, C. A. and Thomson, R. M. (1970). *Physics of solids* (2nd ed.). McGraw-Hill, New York.

Chapter 5

Amberg, C. H. (1967). In *The solid–gas interface* (ed. E. A. Flood), Vol. 2, p. 869. Arnold, London.
Bassett, D. W. (1973). In *Surface and defect properties of solids* (eds M. W. Roberts and J. M. Thomas), Vol. 2, The Chemical Society, London.
— and Parsley, M. J. (1970). *J. Phys.* **D3**, 707.
Blakely, J. M. (1973). *Introduction to the properties of crystal surfaces.* Pergamon Press, Oxford.
Goodman, R. M. and Somorjai, G. A. (1970). *J. chem. Phys.* **52**, 6325.
Henrion, J. and Rhead, G. E. (1972). *Surf. Sci.* **29**, 20.
Ibach, H. (1972). *Proceedings of the 1st International Conference on solid surfaces*, p. 713. American Vacuum Society, New York.
Kaplan, R. and Somorjai, G. A. (1971). *Solid state Commun.* **9**, 505.
Kittel, C. (1967). *Introduction to solid state physics.* Wiley, New York.
MacRae, A. U. (1964). *Surf. Sci.* **2**, 52.

Rosenberg, H. M. (1974). *The solid state* (OPS 9). Clarendon Press, Oxford.
Tong, S. Y., Rhodin, T. N., and Ignatiev, A. (1973). *Phys. Rev.* B8, 906.

Chapter 6

Bassett, D. W. (1973). In *Surfaces and defect properties of solids* (eds M. W. Roberts and J. M. Thomas), Vol. 2, p. 34. The Chemical Society, London.
Bauer, E. and Poppa, H. (1972). *Thin Solid Films* **12,** 167.
Blakely, J. M. (1973). *Introduction to the properties of crystal surfaces.* Pergamon Press, Oxford.
Bond, C. G. (1974). *Heterogeneous catalysis, principles and applications* (OCS 18). Clarendon Press, Oxford.
Brennan, D. and Graham, M. J. (1966). *Discuss. Faraday Soc.* **41**, 95.
Carroll, C. E. and May, J. W. (1972). *Surf. Sci.* **29**, 60 and 85.
Chopra, K. L. (1969). *Thin film phenomena.* McGraw-Hill, New York.
Ertl, G. and Koch, J. (1972). In *Adsorption – desorption phenomena* (ed. F. Ricca), p. 345. Academic Press, London.
Frenkel, J. (1946). *Kinetic theory of liquids.* Clarendon Press, Oxford.
Garmon, L. B. and Lawless, K. R. (1970). In *Structure et proprieties des surfaces des solides*, p. 61. Editions du Centre National de la Recherche Scientifique, Paris.
Gomer, R. (1967). In *Fundamentals of gas–surface interactions* (eds H. Saltsburg, J. N. Smith, and M. Rogers). Academic Press, New York.
Joyner, R. W. and Somorjai, B. A. (1973). In *Surface and defect properties of solids* (eds M. W. Roberts and J. M. Thomas), Vol. 2, p. 1. The Chemical Society, London.
Kittel, C. (1967). *Introduction to solid state physics.* Wiley, New York.
Lord, D. G. and Prutton, M. (1974). *Thin solid films* **21**, 341.
Mayer, H. (1971). In *Advances in epitaxy and endotaxy* (eds H. G. Schneider and V. Ruth), p. 63. VEB Deutsche Verlag für Grundstoffindustrie, Leipzig.
Pashley, D. W. (1970). In *Recent progress in surface science* (eds J. F. Danielli, A. C. Riddiford, and M. Rosenberg), Vol. 3, p. 23. Academic Press, New York.
Pendry, J. B. (1974). *Low energy electron diffraction.* Academic Press, London.
Plummer, E. W. and Bell, A. E. (1972). *Proc. International Conference on solid surfaces*, p. 583. American Vacuum Society, New York.
— and Young, R. D. (1970). *Phys. Rev.* B**1**, 2088.
Prutton, M. (1971). Low energy electron diffraction. *Metallurgical Reviews*, p. 57. Institute of Metals, London.
Sandejas, J. S. and Hudson, J. B. (1967). In *Fundamentals of gas–surface interactions* (eds H. Saltsburg, J. N. Smith, and M. Rogers). Academic Press, New York.
Schrieffer, J. R. (1972). *J. Vac. Sci. Technol.* **9**, 561.
Somorjai, G. A. (1972). *Principles of surface chemistry.* Prentice-Hall, New Jersey.
Vermaak, J. S. and Henning, J. A. O. (1970). *Phil. Mag.* **22**, 269.

Index

Physical constants and conversion factors

Avogadro constant	L or N_A	$6{\cdot}022 \times 10^{23}$ $\mathrm{mol^{-1}}$
Bohr magneton	μ_B or β	$9{\cdot}274 \times 10^{-24}$ $\mathrm{J\ T^{-1}}$
Bohr radius	a_0	$5{\cdot}292 \times 10^{-11}$ m
Boltzmann constant	k	$1{\cdot}381 \times 10^{-23}$ $\mathrm{J\ K^{-1}}$
charge of an electron	e	$-1{\cdot}602 \times 10^{-19}$ C
Compton wavelength of electron	$\lambda_c = h/m_e c$	$= 2{\cdot}426 \times 10^{-12}$ m
Faraday constant	F	$9{\cdot}649 \times 10^{4}$ $\mathrm{C\ mol^{-1}}$
fine structure constant	$\alpha = \mu_0 e^2 c/2h$	$= 7{\cdot}297 \times 10^{-3}$ ($\alpha^{-1} = 137{\cdot}0$)
gas constant	R	$8{\cdot}314$ $\mathrm{J\ K^{-1}\ mol^{-1}}$
gravitational constant	G	$6{\cdot}673 \times 10^{-11}$ $\mathrm{N\ m^2\ kg^{-2}}$
nuclear magneton	μ_N	$5{\cdot}051 \times 10^{-27}$ $\mathrm{J\ T^{-1}}$
permeability of a vacuum	μ_0	$4\pi \times 10^{-7}$ $\mathrm{H\ m^{-1}}$ exactly
permittivity of a vacuum	ϵ_0	$8{\cdot}854 \times 10^{-12}$ $\mathrm{F\ m^{-1}}$ ($1/4\pi\epsilon_0 = 8{\cdot}988 \times 10^{9}$ $\mathrm{m\ F^{-1}}$)
Planck constant	h	$6{\cdot}626 \times 10^{-34}$ J s
(Planck constant)/2π	$\hbar$	$1{\cdot}055 \times 10^{-34}$ J s $= 6{\cdot}582 \times 10^{-16}$ eV s
rest mass of electron	m_e	$9{\cdot}110 \times 10^{-31}$ kg $= 0{\cdot}511$ MeV/c^2
rest mass of proton	m_p	$1{\cdot}673 \times 10^{-27}$ kg $= 938{\cdot}3$ MeV/c^2
Rydberg constant	$R_\infty = \mu_0^2 m_e e^4 c^3/8h^3$	$= 1{\cdot}097 \times 10^{7}$ $\mathrm{m^{-1}}$
speed of light in a vacuum	c	$2{\cdot}998 \times 10^{8}$ $\mathrm{m\ s^{-1}}$
Stefan–Boltzmann constant	$\sigma = 2\pi^5 k/15h^3 c^2$	$= 5{\cdot}670 \times 10^{-8}$ $\mathrm{W\ m^{-2}\ K^{-4}}$
unified atomic mass unit (^{12}C)	u	$1{\cdot}661 \times 10^{-27}$ kg $= 931{\cdot}5$ MeV/c^2
wavelength of a 1 eV photon		$1{\cdot}243 \times 10^{-6}$ m

1 Å = 10^{-10} m; 1 dyne = 10^{-5} N; 1 gauss (G) = 10^{-4} tesla (T);
0° C = 273·15 K; 1 curie (Ci) = $3{\cdot}7 \times 10^{10}$ $\mathrm{s^{-1}}$;
1 J = 10^{7} erg = $6{\cdot}241 \times 10^{18}$ eV; 1 eV = $1{\cdot}602 \times 10^{-19}$ J; 1 $\mathrm{cal_{th}}$ = 4·184 J;
ln 10 = 2·303; ln x = 2·303 log x; e = 2·718; log e = 0·4343; π = 3·142